# SUSTAINABLE ENERGY SOLUTIONS
## for climate change

MARK DIESENDORF is Associate Professor and Deputy Director of the Institute of Environmental Studies at the University of New South Wales in Sydney, Australia. He is researching scenarios for 100 per cent renewable electricity for Australia. His previous books include *Greenhouse Solutions with Sustainable Energy* (2007) and *Climate Action: A campaign manual for greenhouse solutions* (2009).

# SUSTAINABLE ENERGY SOLUTIONS
## for climate change

Mark Diesendorf

Routledge
Taylor & Francis Group
LONDON AND NEW YORK

from Routledge

This edition published 2014
by Routledge
2 Park Square, Milton Park, Abingdon, Oxon, OX14 4RN

Simultaneously published in the USA and Canada
by Routledge
711 Third Avenue, New York, NY 10017
*Routledge is an imprint of the Taylor & Francis Group, an informa business*

and in Australia and New Zealand
By University of New South Wales Press Ltd
University of New South Wales, Sydney NSW 2052, Australia

*British Library Cataloguing in Publication Data*
A catalogue record for this book is available from the British Library

*Library of Congress Cataloging-in-Publication Data*
Sustainable energy solutions for climate change/Mark Diesendorf.
  pages cm
  "Simultaneously published in the USA and Canada"–Title page verso.
  Includes bibliographical references and index.
  1. Renewable energy sources. 2. Renewable energy sources–Government policy.
  3. Climate change mitigation. 4. Climate change mitigation–Government policy.
  5. Sustainability. 6. Sustainability–Government policy. I. Title.
  TJ808.D54 2014
  333.79′4–dc23                                                    2013043474

ISBN13: 978-0-415-70609-4 (hbk)
ISBN13: 978-0-415-70614-8 (pbk)

# CONTENTS

# FOREWORD

This is a handbook for a better future. It reminds us that the unprecedented economic development of the 20th century was fuelled by plentiful cheap energy. The world is now completely different. Analysts established more than 50 years ago that conventional oil production would peak early in the 21st century. The energy source which now powers almost all our transport will become steadily more expensive. Supply interruptions are very likely. So the near-term future will require a new approach to transport. The second challenge for future energy use is global climate change. Our burning of fossil fuels is causing serious problems for all human societies as well as drastic consequences for the natural world. The Millennium Assessment Report warned that we are losing species at an accelerating rate as the driving forces of habitat loss, introduced species and chemical pollution are supplemented by climate change. The report stated that we could lose between 10 and 30 per cent of all mammal, bird and amphibian species this century. These are alarming consequences that demand a concerted international response.

At the same time, the global economic system is still showing the impacts of the 2008 Global Financial Crisis. The World Economic Forum has observed that the problems of fuel, food and finance are three 'canaries in the mine', indicating that the current economic system is simply not sustainable. We are now seeing the outcomes modelled for the Club of Rome over 40 years ago. Their report *Limits to Growth* projected that continuing existing growth trends would see environmental, social and economic collapse in the early to middle decades of this century. As the reports in the United Nations series on the Global Environmental Outlook have been warning for 15 years, the present approach is not sustainable, so doing nothing is not an option.

This book shows that there are realistic and cost-effective solutions. We can move rapidly to renewable energy supply systems. We can also improve dramatically the efficiency of turning energy into goods and services. We live at a level of material comfort that our grandparents could only dream about, made possible by enormous energy flows. Australian energy use is equivalent to us each consuming about 6 kilowatts continuously, about the energy that would be needed for every one of us to drive a small car 24 hours a day. Energy doesn't just light our homes, heat our water and propel our transport vehicles; it is a vital input to providing our food, our drinking water, our dwellings, our clothes and every other aspect of modern life. We don't actually need energy itself; as Amory Lovins said, we don't want energy, we want hot showers and cold beer! We demand the material comfort that is provided by the use of energy.

The technical and political challenge is to find ways of providing our material needs without depleting irreplaceable resources, polluting the air and changing the global climate. This book shows that a clean future is technically and economically achievable. It is possible to live at the same level of material comfort as we do now, using half as much energy or less. We can get all that energy from a mix of renewable energy supply technologies, using the resources of direct and indirect solar energy that will not be exhausted for billions of years. Reducing the environmental impacts of our energy use is only one of the urgent changes we must make, but it is the critical first step because energy provides most of our needs. It is also a significant step on the path to a future that could be genuinely sustainable.

The future is not somewhere we are going, but something we are all creating. At any given time, there are many possible futures. From that wide range, we must be trying to shape a future that is sustainable, at least in principle. At the moment we are fulfilling our desires in ways that reduce options for future generations, by depleting resources. We are essentially stealing from our own descendants. We are also meeting our needs in ways that radically change the global climate and precipitate a collapse of human civilisation. It is criminally irresponsible, but that is the approach we are now following. This book shows that a better future is possible. It is a call to action and a guide for responsible living.

Professor Ian Lowe AO FTSE
Emeritus Professor, School of Science, Griffith University
President, Australian Conservation Foundation

# ACKNOWLEDGEMENTS

I thank John Diesendorf, Brian Martin and my colleagues at the University of New South Wales, Ben Elliston and Iain MacGill, for valuable comments on sections of the manuscript. I also thank Johannes Luetz for unearthing some elusive references; Franziska Mey for providing information and references on energy cooperatives in Germany; Peter Newman for sending me preprints of some of his new articles on transport and urban form; and Frank Stilwell for clarifying some basic concepts in economics. The content of the book has been enhanced by these valuable inputs. However, the views expressed in the book, errors and omissions are my own responsibility.

I am grateful to Khanam Virjee, Commissioning Editor at Routledge-Earthscan, for bringing my idea for the book to her publishing company. At NewSouth Publishing it was a pleasure to work with Heather Cam, Jane McCredie, Elspeth Menzies and Uthpala Gunethilake. Copy-editor Jessica Perini made many insightful suggestions for improving the content and writing of the manuscript. Yvonne Lee transformed several of my rough diagrams into professional figures. Sue Midgley was consistently helpful in liaising on my behalf with several contributors.

Thanks too to Ben Elliston for Figure 2.4; to ACS Cobra/ESTELA <www.estelasolar.eu> for the image of Andasol 1 solar power station (Figure 2.5a) and permission to publish it; to Torresol Energy for permission to publish the image of Gemasolar solar power station (Figure 2.5b); to Mark Cooper for Figure 6.2; and to Peter Newman for Figure 7.1.

I dedicate this book to my sons Thor, Danny and Joey, and my grandchildren Bede, Asha and Oliver.

# INTRODUCTION

In the last ten years ... the most powerful and technically
advanced society in human history ... has been confronted
by a series of ominous, seemingly intractable crises. First
there was the threat to environmental survival; then was the
apparent shortage of energy; and now there is the unexpected
decline of the economy. These are usually regarded as
separate afflictions, each to be solved in its own terms:
environmental degradation by pollution controls; the energy
crisis by finding new sources of energy and new ways of
conserving it; the economic crisis by manipulating prices,
taxes and interest rates.

Barry Commoner[1]

Over the past million years our planet has provided excellent
environmental conditions for nurturing the emergence of human
beings and the development of human societies. But, as our societies evolved from hunter-gatherer to industrial, we have increasingly damaged the system upon which we are totally dependent,
the biosphere, comprising all life on Earth and its non-living environment. We have impacted severely on air, land, waterways, other
species and our fellow humans.

Arguably the greatest and most destructive change is what we
have wrought on the climate. The emission of greenhouse gases
(GHGs) from burning fossil fuels, clearing forests and imposing
destructive agricultural systems has placed us on a trajectory that
could heat this planet to 4 degrees Celsius (°C) or more above the
pre-industrial average by 2100 and drive big changes to precipi-

tation patterns. Impacts include more frequent and more severe droughts, heatwaves, wild fires and floods; rising sea levels damaging coastal infrastructure; loss of biological diversity, including decimation of marine life by acidification of the oceans and the bleaching of coral reefs; declining global food production; and possibly an increase in the frequency of severe storms. Most of these impacts are already being observed.[2]

In addition, local and regional air pollution is a serious environmental and health impact of fossil fuels.[3] You might think filters to reduce air pollution emissions from power stations and catalytic converters to reduce emissions from motor vehicles will go some way to addressing the problem. But the technological improvements are offset by the huge growth in the number of power stations and motor vehicles. Land degradation from open-cut coal mines and mountain top removal is severe.[4] Even underground coal mining can have devastating environmental impacts on the surface, for example where longwall mining causes subsidence.[5]

Energy security was already a concern in the 1960s, when the peak in global oil discovery was passed. It became a more serious issue in 1971, when the USA passed its peak in oil extraction. Australia's modest oil production peaked in 2000. Nowadays many major oil producing countries have already passed their production peaks, suggesting that the world peak of production is now imminent and in future the rate of production will be in terminal decline.[6] Meanwhile, oil consumption by the rapidly growing economies, especially China and India, is growing far above their respective domestic productions. As a result, oil prices are likely to rise steeply as economies recover from the Global Financial Crisis (GFC) and growth in consumption resumes. The struggle for the declining reserves of Middle Eastern oil has already imposed huge costs in terms of the lives of people in the region and the economic burden of US military interventions. Recent claims by journalist George Monbiot and others that peak oil is dead[7] are based on their uncritical acceptance of a flawed analysis.[8]

Won't other fossil fuel technologies save us from peak oil?

There is no cheap, easily produced substitute for conventional oil with comparable volumetric energy density (see Table 7.2). With the exception of natural gas and possibly coal seam methane, all fossil fuel substitutes[9] have comparable or higher GHG emissions and other environmental impacts. The environmental hazards and resulting economic impacts of deep water drilling have already been demonstrated in the Gulf of Mexico.[10] Apart from a temporary glut of shale gas in the USA, substitutes for conventional petroleum-based fuels are all more expensive, thus pushing up prices of transport, food, plastics, etc. Furthermore, recent studies suggest that a global peak in coal production could be reached before 2050.[11] So, even if coal-fired power stations with carbon capture and storage eventually enter the market, there may not be sufficient low-cost fuel to operate them.

The post-World War II economic crises alluded to by Barry Commoner have been dwarfed by the continuing GFC of 2008 onwards. At the time of writing, the crisis is still having severe impacts – in terms of unemployment, under-employment and falling wages – in Spain, Portugal, Italy, Greece, Ireland, Iceland and the USA.

Thus, all the global crises identified by Commoner in 1976 have become much worse. This suggests that technological fixes, while necessary, are not sufficient for solving the linked problems of energy, the environment and the economy. Digging deeper, anthropogenic (human-induced) climate change and the other environmental impacts of fossil fuel use are symptoms or outcomes of three fundamental driving forces: population, consumption per person (sometimes called 'affluence') and inappropriate technology.

## Driving forces of climate change

At one level, we can understand the driving forces of human-induced climate change in terms of a simple mathematical identity (see Glossary), proved in Box I.1:

$$CO_2 \text{ emissions} = \textit{Population} \times \textit{Consumption per person} \times \text{Technology impact}$$

Guided by this relationship, we can understand that total emissions from energy generation in the USA are very high because of very high consumption per person, high population (316 million in 2013) and quite a high proportion of fossil fuel in the energy mix. Total emissions from China are very high, despite low average consumption per person, because of a very high population (about 1300 million) and a very high proportion of fossil fuel, especially coal, in the energy mix. Australia's total emissions are much lower than those of the USA and China because of its relatively low population, 23 million in 2013, but Australia's unenviable record-breaking per capita emissions, the highest among industrialised nations, results primarily from its very high use of coal for electricity generation.

So the basic identity helps us resolve the debate about which driver of environmental impact – population, consumption per person or technology choice – is the most important. Depending on the circumstances, any one of these factors can be dominant. In rich countries growth in consumption per person, coupled with polluting technology, is generally dominant, while population growth, where it is occurring, is a secondary driver. In poorer countries population growth, coupled with polluting technology, is generally the main problem, and economic growth among the wealthier elite may be a secondary driver. The identities are also useful because they show that we can reduce $CO_2$ emissions by addressing each of the three driving factors with separate sets of policies: population with the non-coercive policies (Ch 8, 'Ending population growth without coercion'), energy use per person with programs to foster energy conservation and efficient energy use (Ch 4), and technology with programs to foster both efficient energy use and renewable energy (RE) (Chs 2, 4, 5).

## Climate science is robust

As a scientist involved in the public communication of science, as well as in scientific and technological research on energy systems, I often receive packages from members of the public containing plans for miraculous machines that will perpetually generate useful energy from nothing. If I can find the time, I write back explaining that their invention violates the Law of Conservation of Energy, one of the fundamental laws of physics (see Ch 1).

Though science can tell us very little about beauty, love and ethics, it is the best framework for understanding the structure and functions of natural systems. Based on painstaking observation and experiment, scientists have uncovered fundamental laws of nature, used them to make predictions, tested the predictions and thus identified the limits of validity of the laws. For instance theoretical physicist Albert Einstein has shown that the Conservation of Energy must be modified for bodies travelling at speeds close to that of light, for which it becomes the Conservation of Mass-Energy; other scientists have shown that similar modifications are needed on the scale of fundamental particles. But the original law remains valid under normal conditions of human experience. Many scientists spend a great deal of time, effort and ingenuity in seeking alternative ways of interpreting observations. They consider themselves to be genuine sceptics. However, they generally prefer not to spend a lot of time questioning results that are very well established, unless there is clear evidence based on repeatable observations.

As human beings, the vast majority of scientists do not wish to be seen as 'radical' or alarmist. Therefore, in their public statements about a problem, they tend to be cautious and conservative. For instance, in presenting the results of their very detailed and complex models of Earth's climate systems, climate scientists rarely mention that the models only take into account one of the many positive feedback effects (see Glossary) that amplify global warming resulting from the increasing concentration of GHGs in the atmosphere. The following positive feedbacks are already being observed:

- melting of Arctic ice reduces reflection of sunlight from Earth, thus amplifying global warming
- melting of permafrost releases the GHGs methane and $CO_2$, amplifying warming
- warming of the Arctic Ocean releases methane, amplifying warming
- warming soils release $CO_2$, amplifying warming

- global warming increases the prevalence and intensity of wild fires which release $CO_2$, amplifying warming
- the warming atmosphere holds more water vapour, a GHG, which amplifies warming.

At present, due to the lack of quantitative data on a global scale, only the last of these positive feedbacks is included in climate models. Very few negative feedbacks are known.

In the public debate about climate change many people who reject the science and claim to be sceptics are actually deniers of well-established scientific evidence. They cannot offer an alternative interpretation of the data that stands up to scientific scrutiny. A genuine sceptic, who is a member of the public, would not assume that climate models are over-estimating global climate change. At the very least they would have to acknowledge that they could be under-estimating the changes.

There isn't enough space in this book on sustainable energy to critique the many myths being disseminated by climate science deniers. Scientists have examined these myths and refuted them again and again. The refutations are given on the websites of Skeptical Science[12] and Real Climate.[13]

Just as detectives and forensic scientists identify a criminal by a fingerprint left at the scene of a crime, climate scientists have identified the human responsibility for climate change from the 'fingerprint' of independent observations. Some of the elements of this fingerprint are:

- the average warming of Earth's surface, ocean and troposphere (lower atmosphere)
- cooling of the stratosphere
- night-time minimum temperatures rising faster than daytime maxima
- northern winters warming faster than northern summers
- solar radiation constant over past 50 years, apart from the well-known 11-year solar cycle
- land surface warming faster than ocean surface

- high latitudes (especially Arctic and Antarctic) warming faster than tropics.

Climate science deniers do not have a credible alternative mechanism to that of climate science that satisfies these observations. In particular, variations in the energy output of the Sun or volcanic eruptions could never explain them. The Intergovernmental Panel on Climate Change states that during the last 50 years 'the sum of solar and volcanic forcings would likely have produced cooling, not warming'.[14] The evidence is overwhelming that global warming is real, human-induced and continuing. As a consequence of the emerging positive feedbacks and the high rate of GHG emissions, it must be mitigated urgently.

## 'Do the math!'

Since the industrial revolution the atmospheric concentration of $CO_2$ in the atmosphere has increased from around 280 to 400 parts per million or about 43 per cent. As a result the average temperature of the Earth's surface has increased so far by 0.8°C above the pre-industrial level. However, equilibrium has not yet been reached so that, even if emissions could somehow be stopped today, the temperature would continue to rise until it had increased by about 1.3°C, assuming that air pollution (which has a cooling effect) remains constant, or possibly 2.4°C if air pollution decreases.[15] Current emission trends have put the planet on a path towards warming of at least 4°C before the end of the century.[16] The internationally agreed target adopted by the United Nations Climate Change Conference in Copenhagen in 2009 is to keep the average global warming below 2°C, although this will not necessarily avoid dangerous climate change.

How much $CO_2$ could we emit and still stay below the 2°C guideline temperature increase? A team of scientists from the Potsdam Institute for Climate Impact Research and several universities have calculated that there would be a 25 per cent probability of exceeding this level if we could limit the cumulative $CO_2$ emissions over the period 2000–50 to 1000 gigatonnes (Gt), where 1 Gt is 1 billion tonnes.[17] Over the period of 2000–13 nearly half of this budget was already emitted. Assuming a constant emission rate of 36.3 Gt $CO_2$ per year, the remaining budget would be exhausted in 14 years. In the words of author and climate activist Bill McKibben, 'Do the math!'.[18] The above calculation does not take into account future growth in the emission rate or emissions from GHGs other than $CO_2$.

If we burned existing reserves of fossil fuels equivalent to about 2800 Gt of $CO_2$, we would far exceed the budget. Yet several countries are developing large reserves of fossil fuels. China has the highest production of coal in the world; Canada is extracting oil from a large field of tar sands; the USA is developing gas from shale; and Australia, already the world's biggest coal exporter, is planning to greatly expand its existing coal exports and also to export coal seam methane on a large scale. Clearly the need to stop these and other dangerous developments is very urgent.

## Who is responsible for the climate crisis?

To some degree we are all responsible for GHG emissions through our consumption of fossil fuels and the products and services made from them. However, the principal culprits are large energy-intensive corporations (discussed below) and individuals who are affluent, in the sense of having high consumption per person, and are living in a society with greenhouse-intensive technologies[19] and activities. These include burning fossil fuels, living and working in energy-inefficient buildings, logging native forests, and eating cattle and sheep. Nevertheless, we must be careful about blaming individuals, because most people have limited control over their emissions, in the face of existing institutions, cultures and technology choices available to us.

Large corporations are powerful forces in shaping the economy by lobbying governments and fostering consumer demand through advertising. They encourage growth in consumption per person by pandering to human greed and psychological insecurities. They lobby for government policies to increase population in order to provide a cheap pool of labour and to boost specific industries such as housing/property. They also support the dirtiest technologies in terms of GHG emissions, because these are generally the cheapest. They generally oppose measures to include the environmental and health costs of products and services in their prices. In Australia the big GHG emitting industries boasted to then PhD student/interviewer Guy Pearse that they wrote government policy on energy and climate and called themselves the 'Greenhouse Mafia'.[20]

Underlying the three drivers of emissions – population growth, growth in per capita consumption and dirty technology – are of course greed, the lust for wealth and power, culture (including consumerism), an economic system that encourages growth and speculation, and insecurity resulting from poor governance and an unstable economic system. Although we cannot eliminate greed as a basic human characteristic, with sensible policies we can change the culture and economic system that give it encouragement, support and expression on national and international scales. With appropriate policies, we can make it easy for corporations and individuals to do the right thing. A key challenge is to convince governments that the best interests of the people are not always identical with those of large corporations and that governments should act in the interests of the people. At present, corporations that are big emitters of GHGs are spending many millions of dollars on lobbying and media campaigns that deny the science of human-induced climate change[21] and create the false impression that energy efficiency and RE cannot provide the energy needs of an industrial society. This book challenges that false impression.

## Plan of the book

In this book I aim to show that an ecologically sustainable energy future, based on the efficient use of RE, is feasible, even if we were limited to existing commercially available technologies. Furthermore, I present evidence that the transition to a sustainable energy future, the Great Transition, is affordable, taking into account the costs of continuing with fossil fuels. Although ongoing research and development will always be needed to improve technologies and bring down their costs, the principal barriers to the transition are neither scientific, nor technological, nor economic. They are political and cultural, based on unrealistic ideologies of endless economic and population growth and a society based on consumerism. Although the principal thrust of this book is to demonstrate the feasibility of the transformation of the energy system to one that is ecologically sustainable, socially more equitable and affordable, it also outlines some ideas for transforming the economic system and ending population growth.

Part A of this book addresses the basics of energy: energy resources; the fundamental laws of physics relevant to the conversion of energy from one form to another; and scenarios for the necessary transition to sustainable energy futures. It introduces the concept of ecologically sustainable and socially just development and presents a case that the only energy system compatible with this form of development is one based on the efficient use of RE.

Part B details the technologies and systems for energy supply and use that are likely to provide the basis of a sustainable energy future, and a controversial low-carbon technology that is likely to be a diversion from that pathway. It discusses the environmental, social and economic impacts of these technologies.

Part C addresses the policies needed to transform the energy system, that is, to drive the Great Transition, and create a better, more sustainable society; the barriers to change; and the actors or stakeholders who must drive change: governments, corporations and the community at large. Finally the book engages with the challenge of growing the existing social movement that must pressure, facilitate and assist the key stakeholders to take effective action, both by direct tactics and by creating a new culture of ecological, social and economic sustainability.

# PART A:
# BASIC CONCEPTS AND SCENARIOS

# 1: ENERGY AND ITS GREENHOUSE GAS EMISSIONS

> Without energy, nothing would ever change, nothing would ever happen. You might say energy is the ultimate agent of change, the mother of all change agents.
>
> Dave Watson[1]

Solar energy is the life force of our planet: energising plant and animal life, and driving the great natural cycles that bring us oxygen, drinking water, essential nutrients and a nurturing climate. As people have moved from being hunter-gatherers to members of an industrial society their energy use has grown exponentially and fossil fuels have come to play a dominant role. Since the Industrial Revolution they have brought us great benefits, but now is the time to replace them with benign energy sources that can deliver the same or similar services.

For those readers who do not have a scientific or engineering background this chapter sets out the basic physical principles governing the transformation of energy from one form to another. We look at the current pattern of global energy production and consumption and then focus on the electricity industry, which continuously strives to balance supply and demand. Then we identify energy technologies that are ecologically sustainable and, at least to some degree, socially equitable.

# Natural flows of energy

The Sun provides the energy for almost all processes of life on Earth. From the solar radiation incident on the atmosphere, mostly as visible light, about 30 per cent is reflected into space and the remaining 70 per cent is absorbed by the atmosphere, clouds, oceans and land. The absorbed energy is re-emitted as invisible infrared (heat) radiation. Much of it is trapped by greenhouse gases (GHGs) in the atmosphere, keeping our planet's surface at temperatures suitable for life and driving the evaporation of water, the winds and ocean currents.

FIGURE 1.1    **Food chain in nature**

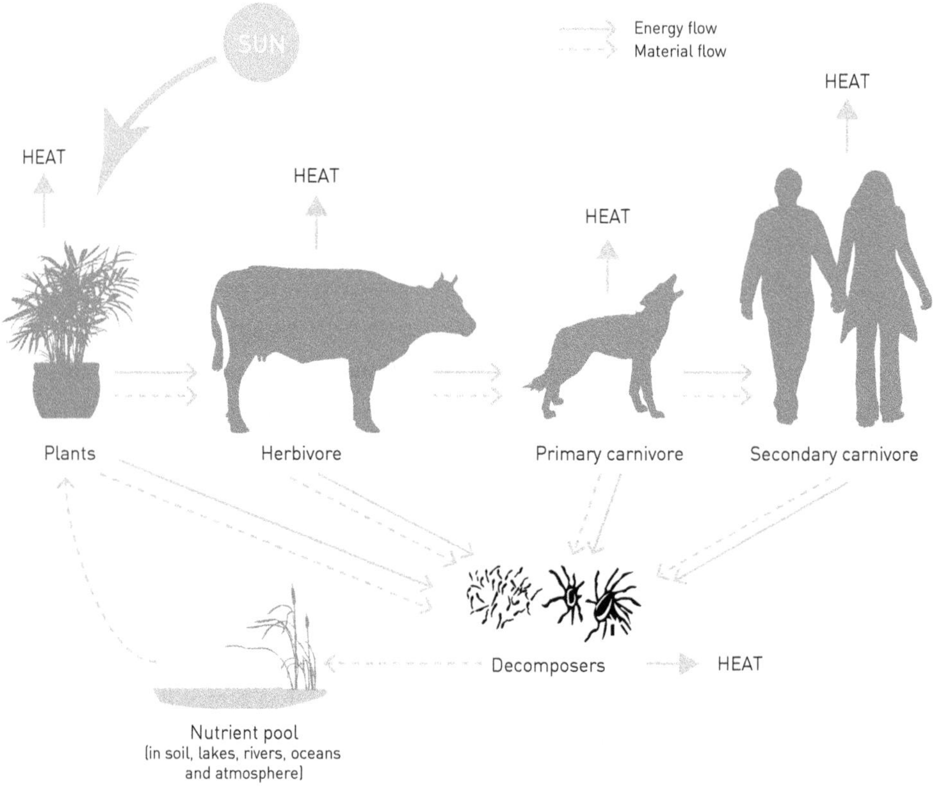

SOURCE Interpretation of Boyden and Dovers (1997, Fig 1.3).[2]

Less than 1 per cent of solar energy reaching the Earth's surface is converted into chemical energy in green plants by the process of photosynthesis. In photosynthesis, carbon dioxide ($CO_2$) from the air, and water from the soil or the sea, react with the Sun's energy to form carbohydrates such as glucose and release oxygen as a by-product. Over half of this process takes place in land plants, while the rest is in the tiny phytoplankton on the surface of the oceans. Plants use some of the carbohydrates they make to release energy for various functions, while the rest is used to increase their bulk (biomass), in a process called respiration, which is chemically similar to burning. In respiration carbohydrates react with oxygen to release energy, $CO_2$ and water. Eventually the carbohydrate energy remaining in the plants is passed on to herbivores that eat the plants, to carnivores that eat the herbivores, and to decomposers – worms, bacteria and fungi – that ultimately eat all species. At each stage in this food chain, energy in the form of waste heat is emitted into the environment (Figure 1.1). The solar energy that is converted from one form to another through metabolic processes within living organisms is called 'somatic energy'.

## Human use of energy

Human ecology is the study of the interactions of humans with other species and their non-living environment. Human ecologist, Stephen Boyden, and environmental historian, Stephen Dovers, identify four phases in human ecology: hunter-gatherer, early farming, early urban and the modern high-energy phase.[3] Table 1.1 summarises the sources of energy in each phase.

With one exception, hunter-gatherers were part of their ecosystems, gaining somatic energy by eating plants and animals, and passing on energy to predators, scavengers, decomposers and the environment. The exception is their use of fire to supplement their own somatic energy. To these sources, early farmers added the somatic energy of animals used for ploughing and transportation. They multiplied their somatic energy by means of simple

TABLE 1.1 **Energy use in four phases of human ecology**

| Phase | Principal primary energy sources |
|---|---|
| 1. Hunter-gatherer | Somatic energy of humans, firewood and other combustible biomass. |
| 2. Early farming | Somatic energy of humans and animals, wood and other combustible biomass. |
| 3. Early urban | Wood, somatic energy, combustible biomass and some wind and hydro. |
| 4. Modern industrial (20th century) | Fossil fuels plus some hydro and uranium supplying a huge increase in energy use over previous stages, typically 50–100 times the rate in hunter-gatherer society. |

SOURCE Boyden and Dovers (1997) plus the author's review of the literature.

tools such as the lever, the block and tackle, and the wheel. Early farmers and town dwellers developed improved technologies to get more use out of fire, for example, to smelt metals. More advanced farmers and early town dwellers converted wind and water power into mechanical energy to grind grain, saw wood and drive simple machinery. Sea transport relied on the wind and human somatic energy. Animals were still used by many farmers in the USA and other industrialised countries as the principal motive power well into the 20th century until they were replaced by tractors.

A huge jump in energy use and associated GHG emissions occurred with the transition to modern industrial society. Initially firewood was the principal fuel for the industrial revolution but, as the forests were felled, coal and then oil and gas took over. The rise of fossil fuels diverted the awareness of many people from the fact that non-commercial natural energy still provides a large contribution to human well-being, through food energy, firewood, timber products and natural heating, cooling and lighting. Human ecologists and environmental scientists remind us that industrial-age humans are still completely dependent upon natural processes, for air, water and food, but live as if they were independent of them.[4]

A detailed history of human energy use, from the Stone Age to the present day, is given in a recent book by physicist Bent Sørensen.[5]

# What are energy and power?

Energy, as discussed in this book, is a concept defined by physicists, in terms of a force moving an object over a specified distance. In the simplest case, energy is the force multiplied by the distance. Since this definition may not be useful to many readers, we focus on the services that energy provides. Energy is the physical driving force of the universe, life and everything. It grows our food and keeps our hearts beating, our homes warm in winter, our food cool and our showers hot. It turns the wheels of industry and the wheels of vehicles that transport us to school, work and holidays.

Energy is classified as either kinetic or potential energy. Kinetic energy is energy of motion, such as falling water, wind, waves, a speeding locomotive, or the rapidly moving molecules of water boiling in a saucepan on a stove. Potential energy is energy stored in some form – such as the energy stored in the chemical bonds in a lump of coal, a torch battery, or the carbohydrates and fats in our bodies. A skier poised on the top of a mountain has potential energy. Energy is measured relative to an arbitrary reference frame: for example, the skier's potential energy relative to the bottom of the chairlift. However, if the skier plans to descend to a hut in a valley below the chairlift, it may be more useful to define her potential energy relative to the hut.

When fossil fuels (or plants) are combusted in the oxygen of the atmosphere, they release energy, $CO_2$ and water, just as in respiration by plants and animals. For plants the amount of $CO_2$ released upon burning is balanced by the net amount (that is, the difference between photosynthesis and respiration) absorbed during their lifetime. For this reason plants are considered neutral in terms of $CO_2$ emissions, *provided* the emission rate from harvesting is balanced by the rate of $CO_2$ absorption during regrowth. If dense forest is cleared and burned and then replaced with a plantation of low density, the balance is destroyed and GHGs are emitted. Of course the combustion of fossil fuels, produced from biomass over periods of up to hundreds of millions of years, results in a huge imbalance between emissions and absorption.

To measure energy most countries use the international SI system of units, in which length is measured in metres (m), mass in kilograms (kg), time in seconds (s) and energy in joules (J). Typical values of the chemical energy stored in a kilogram (kg) of various types of fuel, together with the corresponding $CO_2$ emissions from the combusted fuels, are given in Table 1.2. These values do not include the energy expended in producing the fuel, for example, in converting coal to coke or in refining petroleum. This 'embodied energy' is discussed in Chapter 5, 'Energy payback'. However, in Table 1.2 we are only concerned with the energy that is released when the fuel is combusted. Brown coal has a relatively low energy content because it contains a lot of water and other non-combustibles.

Physicists and engineers call the *rate* of energy conversion

TABLE 1.2    **Energy stored in fuels and corresponding $CO_2$ emissions**

| Fuel | Typical energy content (MJ/kg) | Typical $CO_2$ emissions (kg $CO_2$/GJ)[a] |
|---|---|---|
| Brown coal for electricity | 10 | 90 |
| Black coal for electricity | 23 | 88 |
| Coke | 29 | 120 |
| Dry firewood | 16–20 | [b] |
| Petrol, diesel, kerosene | 46 | 70 |
| Ethanol | 30 | [b] |
| Methanol | 20 | [b] |
| Liquefied natural gas (LNG) at North West Shelf, Australia[d] | 54 | [c] |
| Liquefied petroleum gas (LPG)[d] | 50 | 60 |
| Natural gas | 39 MJ/m³ | 52 |

SOURCE Bureau of Resources and Energy Economics (2012).[6]
NOTES
(a) MJ denotes megajoules = million joules; GJ denotes gigajoules = billion joules.
(b) Net emissions depend on production process.
(c) Value depends upon particular gas field chosen, because of large variations in $CO_2$ content.
(d) Values for LNG and LPG vary slightly with geographic location and date.

TABLE 1.3 **Typical rates of energy use for adults**

| Activity | Power[a] (W) |
|---|---|
| Lying down and resting | 80[b] |
| Hunter-gatherer[c] | 115[c] |
| Walking 4 km/h or cycling 8 km/h or light housework | 220 |
| Walking 6 km/h or heavy housework | 350 |
| Running 16 km/h (4.44 m/s) | 1000 |

SOURCE Based on Boyden and Dovers (1997).
NOTES
(a) The precise rates of energy use of individuals depend on the person's surface area, mass and sex.
(b) Assuming the person has not eaten for several hours. The digestion of food would add about 10 W.
(c) Averaged over both sexes, all ages and a typical day's activities.

TABLE 1.4    **Units of power**

| Unit | Definition | Example from electricity generation |
|---|---|---|
| watt (W) | The basic SI unit of power | Power of a small pilot light is a few watts. |
| kilowatt (kW) | $10^3$ (one thousand) watts | Power output of a single-bar radiator is about 1 kW. |
| megawatt (MW) | $10^6$ (million) watts | Rated power of a large wind turbine is 1–3 MW. |
| gigawatt (GW) | $10^9$ (US billion) watts | Rated power of typical coal or nuclear power station. |

'power'[7] and measure it in watts (W), where 1 watt equals 1 joule per second (1 W = 1 J/s). Tables of units and conversion factors are given at the end of this book. Typical power outputs for human activities that are unaided by fossil fuels are shown in Table 1.3. When scaling up from rates of human energy use to appliances and power stations, we measure power in kilowatts, megawatts and gigawatts, as shown in Table 1.4.

## Energy is conserved

Consider the potential energy stored in the lump of coal, the torch battery, the skier poised on the top of a mountain, and the carbo-hydrates and fats in our bodies. If we burn the coal, switch on the torch, ski down the mountain, or run a mile, we are converting that potential energy into kinetic energy. The amount of potential energy released is equal to the amount stored. Energy is never created or destroyed. That is known as the Law of Conservation of Energy or the First Law of Thermodynamics, one of the fundamental laws of physics.

The *thermal efficiency* of an energy conversion process is defined to be the useful energy output divided by the energy input, expressed as a percentage:

$$Thermal\ efficiency\ (\%) = \frac{useful\ energy\ output \times 100}{energy\ input}$$

(Equation 1.1)

There is a subjective element in this definition, since we decide what is 'useful' energy and what is waste heat. For example, let's assume that a 60-watt incandescent light bulb produces invisible heat energy at the rate of 54 watts (54 J/s). Then its energy efficiency in producing visible light, which we define as the useful

FIGURE 1.2   **Energy flows in generation and transmission of electricity**

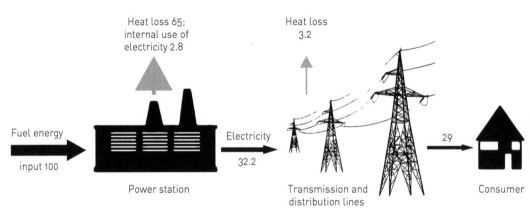

SOURCE The author.

NOTE This diagram is for a typical black coal-fired power station. Given a fuel energy input of 100 (arbitrary units), only about 29 units of energy (that is, 29 per cent of primary energy) reaches the consumer. From a typical brown coal-fired power station, only about 20–25 per cent of primary energy reaches the consumer.

energy output, is 6/60 = 0.1 or 10 per cent. However, if we could also make use of some of the heat emitted by the light bulb, for example, to keep a pet mouse warm, its efficiency would increase.

When black coal is burned in a modern power station, it produces electricity at 35–40 per cent thermal efficiency. The rest is lost as waste heat from the power station. If some of the waste heat from a power station is captured and used to provide heat for nearby industries or buildings, the *useful* energy increases while the fuel use is unchanged and so the efficiency of conversion increases. A power plant that produces both electricity and useful heat together is known as a *cogeneration* plant. In addition, about 7 per cent of the electricity generated is used to operate the power station. (See Figure 1.2.)

To take account of the efficiencies of energy conversion in a black coal-fired power station, transmission and distribution, the overall efficiency of producing light in the bulb is obtained by multiplying together the individual efficiencies expressed as fractions or decimals. Thus, overall efficiency is $0.29 \times 0.1 = 0.029$ or 2.9 per cent. Clearly there is potential for improvement!

## Energy flows 'downhill'

Given that Conservation of Energy is true, what has happened to the energy when coal is burnt to ashes, a torch battery is discharged, a skier stops at the bottom of the slope, and a runner breasts the tape and stops? All the potential energy is ultimately converted to waste heat at low temperature and this is dispersed by the random motion of molecules. For instance, such heat is released as the result of friction experienced by the electrons passing through the filament of the torch light bulb, or the friction of the skis on the snow.

It turns out that, when energy is transformed from one form to another, it tends to go 'downhill' from a higher-grade, more ordered form to a lower-grade, less-ordered form. Examples are: electricity used to heat water; heat flowing from a hot to a cold body; and water flowing downhill. So, energy becomes degraded and less useful to humans. This is one way of stating the Second Law of Thermodynamics. Since entropy is a measure of disorder, another version of the Second Law is that entropy tends to increase.

However, just as it is possible to pump water uphill, against its natural 'inclination', it is also possible to 'pump' energy 'uphill' from low-grade or disordered forms to higher-grade or more ordered forms. This is a very inefficient process, in the sense that much more low-quality energy has to be expended at the input than is received as high-quality energy at the output. Also some high-grade energy is required to drive the process.

For instance, there is a vast quantity of heat stored in the oceans, but it exists at low temperatures, which means that it is low-grade energy. In theory it is possible to generate electricity, a high-grade form of energy, from the small difference in temperature between the ocean surface and the ocean depths as discussed in Chapter 2, but this is a very inefficient process, even in tropical waters where the temperature difference is largest. The thermal efficiency would be – at best – a few per cent. Nevertheless, this may be a viable technology for electricity generation in some parts of the world, if the high-grade energy input driving the process and heating the surface water comes from the Sun, rather than from fossil fuels. If the conversion system is renewable and inexpensive, it may not matter that it is inefficient. After all, green plants are very inefficient at converting solar energy into carbohydrates. But there are lots of plants, so long as we humans conserve them.

Converting one form of high-grade energy (eg, the rotational motion of a turbine) to another form of high-grade energy (eg, electricity) can be achieved with efficiencies of over 95 per cent in both directions. However, because chemical energy is somewhat lower grade than electricity, the conversion efficiency of a conventional coal-fired power station is only about 36 per cent.

# Energy production, consumption and associated emissions

The fossil fuels – coal, oil and gas – are by far the principal sources of energy used in the world today (Figure 1.3). Providing the basis for industrial society was their great contribution. But now, to protect human society and biodiversity, they must be replaced.

FIGURE 1.3    **World primary energy production by fuel, 2010**

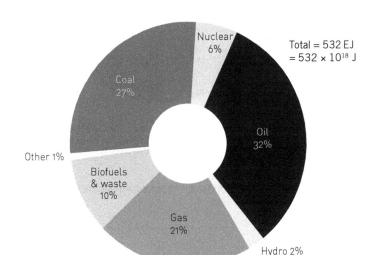

Total = 532 EJ
= 532 × 10^18 J

SOURCE International Energy Agency (IEA) (2012a).[8]
NOTE 'Other' denotes renewable energy (RE) sources other than hydro, biofuels and waste. 'Primary energy' is energy stored in fuels extracted from the environment, before it is converted into other forms such as electricity and gasoline. It includes fossil fuels, uranium and RE sources.

Energy is delivered to users in three principal forms: electricity, fuels to produce non-electrical heat and fuels for transportation (Figure 1.4).

Energy consumption per sector varies greatly from country to country, depending upon the role of heavy industry in the economy and whether transportation by motor vehicle is prevalent. So a global average would not be meaningful. Instead, Figure 1.5 shows

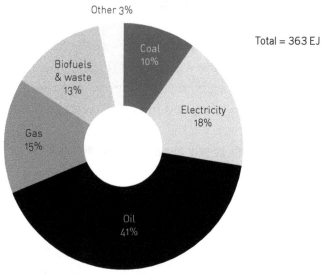

Total = 363 EJ

SOURCE IEA (2012a).[9]
NOTE When going from energy production to consumption, energy has been lost in converting fossil fuels to electricity and in refining crude oil. 'Coal' in this figure denotes the coal remaining after most coal has been converted to electricity; this remaining coal is burned to produce industrial and residential heat.

FIGURE 1.5    **End-use shares of total energy consumption, USA, 2011**

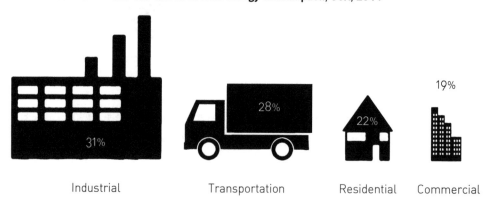

SOURCE US Energy Information Administration (EIA) (2012).[10]

the breakdown of energy consumption in a very industrialised country, the USA. As expected, industry and transportation are the largest energy users, followed by residential and commercial.

Nowadays electricity is used for household appliances, electronic instruments, lighting, air conditioning, some industrial and residential heating, most rail transport and signals. Where coal is widely available – eg, in Poland, South Africa, Australia, China and India – it is the major fuel for electricity generation. It is also the most polluting fuel, both in terms of $CO_2$ emissions and local air pollution.[11] Fortunately electricity is the form of energy that can most easily be generated from renewable sources. Global sources of electricity generation are shown in Figure 1.6.

FIGURE 1.6    **Global electricity generation by fuel, 2010**

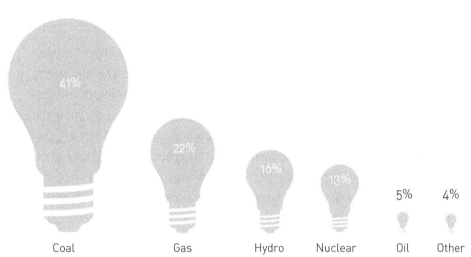

| Coal | Gas | Hydro | Nuclear | Oil | Other |
| 41% | 22% | 16% | 13% | 5% | 4% |

SOURCE IEA (2012a).[12]
NOTE Contributions from nuclear and coal are declining, while those from gas and non-hydro RE ('Other') are increasing.

In regions where it is readily available, gas is the principal fuel for industrial heating and also an important fuel for residential heating. Natural gas is the cleanest of the fossil fuels, both in terms of $CO_2$ emissions and local air pollution. However, recent studies of the life-cycle emissions of gas from shale suggest that it may be more polluting in greenhouse terms than natural gas to the extent that it has little if any environmental advantage over coal.[13] In some less developed countries, coal is still widely used for heat. It is also used everywhere to produce coke, which is both an energy source and a chemical reducing agent for the production of metallurgical steel. While much low-temperature heat (eg, for hot water, space heating and cooling) could be obtained directly from solar energy, to produce high-temperature heat from renewable energy (RE) will be more challenging. In a low-carbon world, it seems likely that most of it will have to come from renewable electricity.

At present the principal fuel for motor vehicles is of course oil. Renewable electricity appears to be a feasible option for urban cars and trucks but, because of the limited range of batteries, it is not an option for long-distance trips between cities or in rural areas, where biofuels could make a contribution.

Thus further electrification is an important feature of sustainable energy scenarios. Therefore transforming the electricity industry to one based on RE must be a priority in climate mitigation. According to the International Energy Agency, in 2010 the generation of electricity and heat combined comprised the biggest single source of global anthropogenic $CO_2$ emissions (41 per cent) followed by transport (22 per cent). About three-quarters of the combined electricity and heat generation came from the combustion of coal. About three-quarters of the transport emissions came from road transport.[14] Since electricity is the key to a sustainable energy future, we focus on the basics of the electricity industry.

## Balancing electricity demand and supply

An electricity supply system is actually a delicate balancing act between supply and demand. As weather, business, industrial output and residential activities vary, demand varies with time of day, day of the week and season of the year. Because the storage of electricity on a large scale is still very expensive, electricity supply in a conventional system has to track demand with a high degree of precision. Batteries that could smooth fluctuations in supply and demand in a national electricity grid are still at the pilot or demonstration stage (see Ch 2, 'Storage of renewable electricity'). Even small deviations between the supply and demand can produce changes in the frequency of the alternating current that can damage electronic equipment. Larger deviations can produce blackouts.

The most challenging task for an electricity supply system is to follow the large and rapid demand variations over a 24-hour period while maintaining reliability. Figure 1.7 shows typical diurnal demand variations, from midnight to midnight, in summer (left-hand side) and winter (right-hand side) in an industrialised country. In summer, generally a single large peak occurs in the late afternoon due to air conditioning. The two largest peaks in winter are due mainly to residential space heating during the periods around breakfast and dinner.

The horizontal lines in Figure 1.7 mark out the boundaries between three categories of demand: base load, intermediate load

FIGURE 1.7   **Typical diurnal electric power demand by time of day**

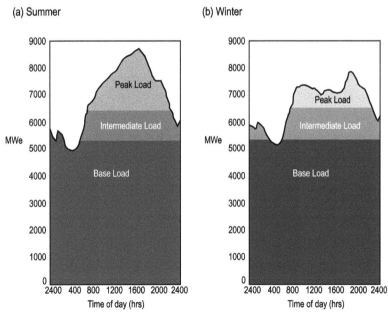

SOURCE The author.

and peak load. The trough in demand between midnight and dawn, projected across the full 24 hours, defines *base-load* level of demand; the big peaks are called *peak-load* demand; and between them is *intermediate load*.

In the past it was appropriate to match each of these demand categories with a specific type of power station. Coal-fired and nuclear power stations are inflexible in terms of varying ('ramping') their outputs rapidly. They can take a day or more to start up from cold and build up to full power. Once they are running, they cannot follow the diurnal variations in demand without incurring great expenses in operation and maintenance. They have high capital costs and, when operated continuously at rated power, low fuel costs and other operating costs. So preferably, they should be operated continuously at their rated (maximum design) power output. Thus it is natural to match them with base-load demand and label them as *base-load power stations*.

To meet the peaks in demand and to help fill the gap in supply when a base-load power station breaks down unexpectedly, *peak-load power stations* are used. These can be started up within minutes and their outputs can be varied or ramped rapidly. There are two principal types of peak-load power station: gas turbines and hydro-electric stations based on storage dams. The gas turbines used for electricity generation are very similar to jet engines on aircraft. They can be brought from cold to full power in about 10 minutes. They have low capital costs in dollars per kilowatt ($/kW) of rated generating capacity. They have high running costs, because the fuels they burn – natural gas, oil or renewable equivalents – are much more expensive than the coal or uranium used in base-load power stations. Thus peak-load power stations are operated as little as possible. They may be considered to be a type of reliability insurance.

Storing water is the principal conventional way of (indirectly) storing electricity on a large scale. Although rainfall varies often substantially by season and year, a large dam can smooth out these natural fluctuations. In the few regions with very large hydro-electric storage (eg, Norway, Iceland, Brazil, New Zealand and Tasmania), hydro can be operated as base-load power. However, in most countries, storage and/or rainfall capacity is limited and hydro is peak load or at best intermediate load. The output of hydro stations with large dams can be varied even more rapidly than that of gas turbines. However, unlike the latter, hydro stations with reservoirs generally have high capital costs. Although their 'fuel', the water they store, is provided free of charge by nature, for storages that are not gigantic this limited resource has to be used sparingly, as if it had a high fuel price.

Intermediate-load power stations, as the name suggests, are intermediate to base and peak in several characteristics. Their output can be varied more quickly than that of base-load stations, but more slowly than peak-load. Their fuel costs generally lie between those of base-load coal and peak-load gas. Their capital costs in \$/kW of rated generating capacity fall between those of base-load and peak-load stations. They are usually combined-cycle gas turbine stations or old, small, coal-fired power stations whose capital costs have been paid off. A combined-cycle power station has two stages – the waste heat from a gas turbine is captured to produce steam in the boiler – and so is more efficient than a single cycle power station. Table 1.5 summarises the properties of conventional base-, intermediate- and peak-load power stations.

TABLE 1.5    **Properties of different types of conventional power station**

| Type | Fuels | Capital cost (annualised) | Operating cost (mostly fuel) | Ability to ramp output | Capacity factor |
|------|-------|---------------------------|------------------------------|------------------------|-----------------|
| Base load | Coal, nuclear | High | Low | Low | High |
| Intermediate load | Coal, gas | Medium | Medium | Medium | Medium |
| Peak load | Gas, hydro, oil | Low (gas/oil); High (hydro) | High | High | Low |

SOURCE The author.

Although a coal-fired power station, and indeed any base-load power station, is designed to be run continuously at its rated power, it does not always do this. Sometimes it breaks down and sometimes it is shut down for planned maintenance. When demand for electricity on the grid is very low, a coal-fired power station may be run at reduced power[15] or shut down. So, the power output averaged over a year is always less than the rated power. We define the *capacity factor* of a power station to be its average power output divided by its rated power, expressed as a percentage.

$$Capacity\ factor\ (\%) = \frac{average\ power \times 100}{rated\ power} \qquad \text{(Equation 1.2)}$$

$$= \frac{annual\ energy\ generation \times 100}{hypothetical\ annual\ energy\ generation\ at\ rated\ power}$$

So, capacity factor is partly a measure of performance of the power station and partly a measure of how it is used by an electricity supply system. For base-load power stations, typical capacity factors, averaged over the plants' lifetimes, are in the 60–90 per cent range. Because of their high fuel costs, peak-load power stations are operated as little as possible, with the result that typical capacity factors are 2–10 per cent. Hence, for peak-load stations (and, we shall see, for most RE systems), capacity factor is not a good measure of efficiency. Intermediate-load power stations, as always, are in the middle.

In an electricity supply system that is based predominantly on fossil and/or nuclear fuels, a combination or mix of base-, intermediate- and peak-load power stations give the minimum annual cost. If there is too much base-load plant, total costs are high, because of the high annual repayments on the capital cost. If there is too much peak-load gas turbine use, total costs are high, because of the high use of the expensive fuel of the peak-load plant. Striking the correct balance between the different types of conventional power station to achieve the optimal economic mix is difficult, especially if average annual demand is changing significantly from year to year. It involves projection and prediction, which can be very inaccurate. The calculation is even more complicated in supply systems with a significant fraction of hydro-electricity.

A conventional electricity supply system is designed to handle two kinds of variability. Predictable variability is the broad shape of the daily and seasonal demand curves and the planned shutdowns of power stations for maintenance. This is handled by having an optimal mix of base-, intermediate-, and peak-load power stations with sufficient total generating capacity to supply predictable maximum demands and to back up planned shutdowns. But an electricity supply–demand system is not completely predictable. Both supply and demand can also vary randomly, through unexpected breakdowns of power stations or transmission lines and unexpected changes in demand, such as may occur when millions of people put on the kettle during an advertising break in a popular TV program or switch on heating or cooling appliances when there is a sudden change in weather. Of course, these events are not completely random, like the toss of an unbiased coin, but they do involve degrees of uncertainty. Power system planners try to calculate the probabilities of such events and budget for additional reserve power stations to handle them with a high degree of certainty.

Even so, no power system is 100 per cent reliable. To achieve that impossible goal would require reserves for the reserves and so on ad infinitum, thus imposing an infinite cost on electricity consumers. So, power system planners generally choose a politically acceptable reliability that involves infrequent failures of supply to meet demand. To satisfy that requirement a common precaution is to have additional peak-load plant in the system and sufficient base-load reserves hot and ready to step in if the largest power station in the system fails unexpectedly. Some cold base-load reserves could also be brought on-line over a period of days if the breakdown appears to be long-term. The duplication of key transmission links is another reliability precaution.

# Enter 'intermittent' renewable energy

Some RE sources are regarded as 'intermittent', 'fluctuating', 'non-dispatchable' or 'variable', while others are regarded as 'not intermittent', 'reliable', 'flexible' or 'dispatchable' (see 'dispatch' in Glossary). In the former group are wind, solar PV without storage, run-of-river hydro and wave, while in the latter group are biomass power stations, hydro with large reservoirs, geothermal, tidal and concentrated solar thermal power with large amounts of thermal storage. It must be emphasised that 'random' fluctuations in 'intermittent' RE sources are partially predictable, just like 'random' fluctuations in demand and conventional supply. Mathematically, all three types of random fluctuation can be described by probability distributions.[16]

When variable types of RE are introduced into the grid, they are initially a tiny fraction of annual electricity generation. Then the fluctuations in output from wind and solar photovoltaic (PV) power are generally smaller than the fluctuations in demand and so can be handled easily by the existing intermediate-, peak-load and reserve base-load stations. Thus the power output from small penetrations into the grid of variable wind and PV can be treated simply as negative demand, that is, equivalent to reductions in electricity demand when the wind is blowing and/or the Sun is shining.

Among electrical engineers a conventional view was that, although these small fluctuations are not a problem, large contributions from variable forms of RE, especially wind and PV, would introduce new, large sources of 'random' variation and so would decrease the reliability of the generating system. Therefore, it was argued by some, large penetrations of wind and PV would either need 100 per cent back-up by conventional power stations, or would have to await the development and commercialisation of base-load renewable electricity technologies, such as geothermal, or low-cost electrical storage.

However, the late 20th and early 21st centuries saw rapid growth of variable RE technologies from a small base to substantial levels in several countries and states. In 2012 wind power supplied 30 per cent of annual electricity generation in Denmark and 27 per cent in South Australia. While there are indeed challenges in handling large penetrations of RE into the electricity grid, they are not as difficult or expensive as previously thought.[17] Both practical experience and detailed computer modelling discussed in Chapter 3 have led to changes in the conceptual framework of an electricity supply system. With smart design, a predominantly renewable electricity system can be just as reliable as one based on fossil fuels. There is no need to supply base-load demand by means of base-load supply. Indeed, in a predominantly RE system there is no need for any kind of base-load supply. As shown by the computer simulations discussed in Chapter 3, the key is to maintain the reliability of the whole supply–demand system at the required level by means of an appropriate mix of 'flexible' and 'variable' RE technologies. Wind power, solar PV and run-of-river hydro cannot be operated if there is no wind, sunshine or water flow, respectively. Therefore they are regarded as 'variable' and 'non-dispatchable', although they can be predicted to some degree, while concentrated solar thermal with thermal storage, hydro with storage, geothermal and gas turbines with a reliable fuel source, whether it be renewable or fossil, are regarded as 'flexible' and 'dispatchable'.

The technologies are described in more detail in Chapters 2 and 5. Here we justify the choice of RE technologies as the supply component of a sustainable energy system.

## Ecologically sustainable energy system

An ecologically sustainable energy system is one that is compatible with, and preferably facilitates, the development of an ecologically sustainable and socially just society. It is essential to the sustainable development process. The best-known definition of sustainable development is from the World Commission on Environment and Development:

Humanity has to make development sustainable – to ensure that it meets the needs of the present, without compromising the ability of future generations to meet their own needs.[18]

This definition emphasises the long-term aspect of sustainability. A complementary definition, that addresses the relationships between ecological, economic and social aspects of the sustainable development process, is:

Sustainable development is economic and social development that protects and enhances the environment and social equity.[19]

This strong definition takes the view that sustainable development is not some kind of balance or trade-off between environmental, economic and social aspects, but rather that environmental protection and social equity should be constraints on the kinds of development permitted. 'Social equity' is defined here as equal opportunity for people to achieve certain basic needs, such as food, shelter, clean air and water, access to basic energy services and personal security.

In this context, a *sustainable energy system* is a system of technologies, laws, institutions, education, industries and prices governing energy demand and supply for the sustainable development process and ultimately for achieving a sustainable society. Clearly this goes far beyond a simple choice of technologies, to a system needing a wide range of policy instruments. Ideally such a system should:

- have very low adverse impacts on environment, health and safety
- have large 'fuel' reserves, while using energy, materials and other resources efficiently
- be affordable for all (a statement about society as well as its energy system)
- be compatible with democratic control of the technology and
- create local employment.

At present, three types of energy technologies are either commercially available or at least at an advanced stage of demonstration: those based on the combustion of fossil fuels; those based on nuclear fission; and those based on RE. Fossil fuels fail the above sustainability criteria on all counts. They are the principal cause of global climate change; a major cause of air pollution with widespread severe impacts on human health; a significant cause of water pollution and land degradation; have very limited reserves (especially oil and good quality coal); are only relatively cheap because their environmental, health, social and some economic costs are excluded from their prices; are controlled by huge corporations; and create relatively few jobs per unit of energy generated. Of the fossil fuels used on a large scale, coal is the most damaging in terms of environmental and health impacts.[20] Also, coal mining is one of the most dangerous occupations, even in industrialised countries.

Nuclear power, discussed in detail in Chapter 6, increases the proliferation of nuclear weapons and hence the risk of nuclear war. It is potentially a high-impact target for terrorism and also exposes large populations to the risk of rare but devastating accidents. It is very expensive and only survives commercially because of huge continuing subsidies. After more than half a century of commercial operation no facility anywhere in the world exists for the long-term management of high-level nuclear wastes. Even escapes of low-level radiation will have a large cumulative health impact over periods in the order of 100 000 years. Although life-cycle $CO_2$ emissions are currently low, this temporary situation depends on the use of high-grade uranium and its reserves are now very limited. If low-grade uranium is mined and milled using fossil fuels, life-cycle $CO_2$ emissions would become a large fraction of the emissions from natural gas. Nuclear power creates very few jobs and is incompatible with democratic control. Thus it also fails all the sustainability criteria.

The adverse impacts of both fossil fuels and nuclear energy generally fall most heavily on the poor and disadvantaged. The industries tend to locate hazardous installations near low-income and minority communities.[21]

Most RE sources satisfy all except possibly the affordability criterion. (Exceptions are some hydro-electric dams and some sources of bioenergy, which can have substantial adverse environmental impacts, including GHG emissions, and adverse social justice impacts if they take land away from food production – see Chapter 5.) Several of the principal renewable technologies – on-shore wind, solar hot water, solar-efficient design of buildings, solar PV modules and geothermal heat pumps – are already affordable in many regions of the world, while the cost of concentrated solar thermal power with thermal storage is expected to be halved as its market expands. The relatively small component-size of most solar and wind systems makes manufacture of components in many countries more feasible than for coal and nuclear power stations. While no energy system is completely benign, it is already clear that those based on the efficient use of RE can satisfy sustainability criteria.

## Conclusion

Industrial society has rates of energy use 50–100 times that of our hunter-gatherer ancestors. These high rates of energy use have enabled a dramatic socio-economic transformation, bringing many benefits. But supplying this energy by burning fossil fuels is rapidly changing Earth's climate, polluting the planet in several other ways, and leaving us dependent upon fuels that are running out. Furthermore, it is impacting more heavily upon the poor and disadvantaged.

The solution is a radical one, namely to transform the existing energy system into one that is ecologically sustainable and more socially just. At present the only technologically proven system that is (mostly) commercially available and meets sustainability criteria is the efficient use of RE, called 'sustainable energy' hereinafter. The remainder of this book addresses the opportunities and challenges of the Great Transition to sustainable energy: the technologies, the scenarios for change, the policies needed and strategies for mobilising the people and governments to accept these policies.

# 2: ENERGY RESOURCES AND TECHNOLOGIES

We are like tenant farmers chopping down the fence
around our house for fuel when we should be using Nature's
inexhaustible sources of energy – sun, wind and tide ... I'd
put my money on the sun and solar energy. What a source of
power! I hope we don't have to wait until oil and coal run out
before we tackle that.

Thomas Alva Edison[1]

While renewable energies (RE) will continue to be available to
Earth's inhabitants until our Sun eventually dies billions of years
in the future, their flows are more diffuse, in terms of watts per
square metre, than those of fossil and nuclear fuels. So the question
arises, is there enough RE reaching our planet's surface to fulfil the
present and future needs of human civilisation? To what extent
would there be competition for land between capturing sunlight
for food production and energy production? If there is an abun-
dance of RE, do we have the technologies for converting it to run
our homes, industries, transport and other activities?

To address these questions and prepare the ground for the discussion of sustainable energy scenarios in Chapter 3, this chapter
offers a concise overview of global RE resources and the technologies for converting these resources into commercial energy. Of
particular interest is the status of the diverse technologies in terms
of stage of development or maturity. First we consider resources.

# Renewable energy resources

The rate of solar energy input (insolation – not to be confused with insulation) to the Earth could supply approximately 6700 times the rate of energy demand of an industrial society. The simple calculation is as follows. Averaging over the planet, over night and day, and over seasons, the rate at which solar energy reaches Earth's surface is about 200 watts per square metre ($W/m^2$). For comparison, the current rate of global primary energy consumption is about 18 terawatts (TW); dividing by the surface area of the Earth gives an average energy density of 0.030 $W/m^2$. Dividing 200 by 0.03 gives 6666.

Even after allowing for land occupied by forests, farms and cities, there is land to spare. For simplicity in an approximate calculation, let's assume that in future all energy will be used in the form of electricity. Allowing an efficiency of energy conversion of 20 per cent and spacing the solar collectors sufficiently so that they don't shade one another gives a total land area of about 500 000 square km, equivalent to a square of side 700 km. This could be broken up into smaller areas situated in the desert regions of each inhabited continent, except Europe which could be energised from North Africa. About one-quarter of the area would be on roofs of homes, factories, warehouses and parking lots.

In practice a 100 per cent RE system would comprise a mix of different RE sources, not only direct solar. Wind is produced by the uneven heating of the land, sea and atmosphere by the Sun and so is an indirect form of solar energy. On-shore wind turbines are generally placed on agricultural land where they, together with access roads and substations, occupy only 1–3 per cent of the land area. Waves (except tsunamis) are caused by wind blowing across the water. Like off-shore wind farms, wave power stations occupy negligible land and do not compete with food production. Geothermal heat and power also has low demands on land.

The only technology that could occupy land to the extent of competing with food production is bioenergy based on dedicated crops. As a minor component of electricity generation the modest amounts of bioenergy available from the residues of existing crops and forests would suffice in most regions. However, biomass residues would not provide an adequate supply if liquid and gaseous biofuels are used extensively for motor vehicles, aircraft, heating and the chemical industry, as well as for electricity generation. For this reason, many of the RE scenarios discussed in the next chapter have quite small contributions from bioenergy. If hydrogen could be produced economically by using RE, the pressure to grow dedicated energy crops on a large scale would subside. But a large-scale hydrogen 'economy' is not close (see Ch 7, 'Improved vehicles and fuels').

Given that competition for land is unlikely to be a major problem for most solar, wind, marine and geothermal installations, we are now ready to address the question of whether there are technologies available to convert these abundant sources into usable energy.

## Maturity of energy technologies

A common misconception about RE technologies, disseminated by its opponents, is that they need much more research and development (R&D) before they can be widely used. Until that work is done, the RE deniers claim, we need coal with carbon capture and storage (CCS) and/or nuclear power during the allegedly long transitional period to an RE future.

However, reality is the opposite of this. Coal power with CCS is unproven as a whole technological system, although several individual components of the system are at the demonstration stage. At the time of writing there is no large-scale operating coal-fired power station with CCS, the most challenging project, although there are a number of large-scale systems that capture $CO_2$ from oil and gas fields and industrial plants.[2] Coal power with CCS still needs much research, development and demonstration before it can be considered seriously as a significant option. However, progress has stalled as an awareness of the magnitude of the technological challenge and its costs has grown. Recent estimates for both capital and operating costs of coal power with post-combustion CCS systems are approximately double those of conventional coal-fired power costs.[3] These estimates exclude the costs of transporting and storing the $CO_2$ after capture. Costs can be substantial for some locations of power stations, such as New South Wales.

The International Energy Agency's (IEA's) principal energy scenario, known as the Blue Map Scenario, has coal power with CCS making a negligible contribution to global energy supply in 2025, a very small contribution in 2030 and only overtaking RE around 2050. It also has new nuclear supplying a tiny fraction of the new RE contribution in 2025 and only reaching one-third of the RE contribution in 2050.[4] That is the result of the very long planning and construction time needed by nuclear power stations and associated infrastructure.[5] While Blue Map is a scenario rather than a forecast, it is the IEA scenario most favourable to CCS, reflecting the IEA's understanding that coal power with CCS is still in the early stages of development.

New technologies generally go through an innovation process with different stages – from research to development to demonstration to early diffusion to commercial maturity, as set out in Table 2.1. This classification is useful, because technologies at one stage of the process need a different set of policies to progress them than technologies at other stages. More precisely, each stage needs emphasis on different policies. For example, R&D play a vital role at the beginning of the innovation process, but for technologies already in limited or full-scale mass production, R&D have a minor role compared with policies to expand the market, discussed in more detail in Chapter 9.

During the *R&D* stages the technology undergoes laboratory study and/or small-scale study in the field. The main focus is on proving the basic concept. This stage is far from giving any credible cost estimates. Examples of current R&D on potential low-carbon sources of energy and related technologies are novel pathways for

TABLE 2.1    **Stages of maturity of low-carbon energy technologies**

| Stage | Description | Technologies |
|---|---|---|
| Commercially available and mature | Mass production of well-tested systems at an accurately predictable price, aiming for a large market. The need for a carbon price does not invalidate this classification. | Energy conservation; many energy efficiency (EE) technologies and designs for buildings, appliances, stationary equipment and motor vehicles; passive solar house design; urban consolidation around public transport routes and nodes; solar hot water; on-shore wind power; landfill gas; biomass co-firing with coal; biomass combustion for generation and cogeneration; ethanol from fermentation of sugars; solar photovoltaics (PVs) based on silicon crystal and thin film collectors based on cadmium telluride; 2nd generation nuclear power; hybrid vehicles; cogeneration and trigeneration; conventional geothermal power; geothermal heat pumps for heating and cooling large buildings; tidal power with dam; pumped hydro storage; bicycle; heavy rail and light rail; lithium-ion and nickel–metal hydride batteries for electric vehicles (EVs). |
| Early diffusion | Limited mass production of several large-scale systems having designs whose main features are well understood; some detailed design features and production methods still being optimised in preparation for large-scale mass production. | Off-shore wind power; concentrated solar thermal (CST) with parabolic trough collectors and molten salt thermal storage; compressed air storage; 3rd generation nuclear power; trigeneration fuelled by gas or biomass; plug-in hybrid and all-EVs; vanadium redox batteries; fuel cells for hydrogen and methanol; solar space heating and cooling. |
| Demonstration | Field deployment of a single unit or very few units of the technology. Unit is large, although rarely full size. Designed with mass production in mind, but not all main features are fully understood at this scale. | CST with central receiver; CST linear Fresnel; CST big dish; biofuels from ligno-cellulose; cogeneration and trigeneration based on CST; CCS at gas field; wave and ocean current power; UltraBattery; thermal energy storage based on graphite blocks and ammonia dissociation; enhanced/engineered geothermal power; fast breeder reactor; some components of CCS systems. |
| R&D | Undergoing laboratory or small-scale field study. Main focus on proof of concept. Design is not directed towards future production processes. | Novel PVs; artificial photosynthesis; biofuels from ligno-cellulose; hydrogen production by microbial action on biomass; liquid fuels from algae; coal-fired power with CCS as a whole system; integral fast reactor; controlled nuclear fusion. |

SOURCES Adapted from United Nations Framework Convention on Climate Change (UNFCCC) (2008)[6] and Foxon et al. (2005).[7]

producing hydrogen from genetically engineered organisms; low-cost processes for converting biomass (organic material) to liquid or gaseous fuels, including liquid fuels from algae; artificial photosynthesis; new low-cost solar photovoltaic (PV) cells; CCS from existing coal-fired power stations; and controlled nuclear fusion (which has been stuck at this stage for half a century).

Technology that survives the R&D stages may enter the *demonstration* stage. This involves deployment of a few units of the technology in the field. Unit size is often of a medium scale to keep costs down. Much learning takes place at this stage, since scaling up from laboratory or pilot scale to medium scale frequently reveals engineering problems that must be overcome if the technology is to advance. The demonstration units are designed with future mass production in mind, on the understanding that operational experience is likely to suggest significant design modifications, especially for improving reliability and facilitating potential future manufacturing processes. Successful demonstration should permit approximate estimates of future costs in mass production and so may awaken the interest of financial institutions into keeping a watching brief for possible future investment. Examples of energy technologies at the demonstration stage at the time of publication are:

- several types of wave and ocean current power technologies
- some production processes for second generation biofuels from ligno-cellulose
- concentrated solar thermal (CST) power based on central receiver systems or paraboloidal dishes or linear Fresnel collectors
- thermal energy storage based on graphite blocks or the dissociation of ammonia
- the UltraBattery (a combination of capacitors and lead–acid battery)
- utility-scale energy storage systems
- hot rock or engineered geothermal power
- CCS at gas fields and
- fast breeder reactors.

*Early diffusion* is self-explanatory. The technology is commercially available and full-size units are deployed in the field to a limited degree. Mass production of components occurs on a limited scale. Since the market is small, the price is much higher than it would be in large-scale mass production. Production methods are still being optimised. Examples include:

- off-shore wind turbines
- CST power stations based on parabolic trough collectors
- geothermal heating and cooling of buildings
- electricity and heat from the gasification of biomass and
- plug-in hybrid and electric vehicles.

Third generation nuclear power stations are also placed in this category, because very few are operating as yet, although they are similar to second generation, the commonest types of nuclear power station.

In contrast to the technological status of coal power with CCS, many key sustainable energy technologies are *commercially mature*, meaning that they are in large-scale mass production and can be ordered at a predictable price. However, to be classified as commercially mature, they do not have to be economically competitive with conventional energy technologies, since the latter are generally not priced appropriately (given that high subsidies are not accounted for, nor the environmental and health damage they cause). At this stage the market continues to grow, so there is still potential for declining prices. For efficient energy use there is a wide range of commercially available technologies, including insulation, efficient designs for appliances and equipment, and several basic designs for solar-efficient homes and commercial buildings suitable for diverse climate zones. Examples of commercially mature RE supply technologies are hydro-electricity, with or without pumped storage; tidal power based on damming estuaries; efficient wood-burning space heaters; solar hot water; solar PV modules based on silicon crystals or some types of thin film; on-shore wind power; electricity from landfill gas; ground-source ('geothermal') heat pumps;

**Turbines**

Physics tells us that rotating a coil of conducting wire in a magnetic field generates an electric current. The kinetic energy of rotation is converted into kinetic energy of moving electrons making an electric current. This is the principle of the generator, which can be designed either to produce direct current (DC) or alternating current (AC). The generator is driven by a turbine, which converts the incoming linear motion or gas expansion into rotation. There are several different types of turbine, each designed to handle the incoming source of energy best.

- A water turbine converts a linear flow of water into rotation, whether that flow is the result of gravity in a hydro scheme or a tidal/ocean current. The turbine shapes are very different for high-speed hydro and low-speed ocean current.
- A steam turbine is rotated by a jet of steam from boiling water, which may be produced by burning a fossil or renewable fuel, or by the heat released by nuclear reactions.
- For electricity generation, wind turbine blades are similar in shape to aircraft wings – they fly on the wind. In contrast, the turbine of a water pumping windmill is just a fan that is pushed by the wind.
- A gas turbine, such as a jet engine, is rotated by burning a liquid or gaseous fuel, fossil or renewable, in air. The gaseous mixture expands and flows through a nozzle to hit the turbine blades.

The descriptions of operation of the steam and gas turbines have been simplified for the general reader. If you wish to learn about the complex processes involved and the detailed science and engineering behind them, read Bent Sørensen's *Renewable Energy*.[8]

conventional geothermal power; hybrid cars and the lithium-ion batteries they use; biomass co-firing with coal in power stations; and gas turbines that can burn gas and liquids from both fossil and renewable sources. A non-renewable technology, Generation II nuclear power, is also generally regarded as commercially mature because so many have been built, although nuclear power stations are rarely completed on time and on budget (see Ch 6).

The development of technologies does not always follow a linear process from R&D to commercially mature. At the demonstration and early diffusion stages, there may be several different designs competing for the same task, several generations of a single design competing with one another, and dead-ends and loops in the process. Furthermore, the divisions between the various stages are not sharp. There is debate about which technology belongs in which stage.

In the following sections I outline the resources and technological features of the key technologies in the RE scenarios discussed in Chapter 3. Since almost all renewable electricity technologies generate electricity by means of a turbine, this key mechanical component is discussed in Box 2.1.

## Solar energy

Solar energy can be used to heat water, buildings and industrial processes; to produce steam for electricity generation; and to produce electricity directly in a solar PV cell. These pathways are illustrated in Figure 2.1. In this section I outline the different solar technologies, including collectors and storage, their current status and future prospects.

### Collection of solar energy

The solar collectors used for passive solar housing, solar hot water and rooftop solar PV are flat and so can accept incoming solar energy from all directions, as occurs on an overcast day when the Sun is invisible but the cloud is thin enough to still give a fairly bright sky. For such solar collectors, the appropriate solar data is measured by the sunshine falling upon a flat horizontal plate and

FIGURE 2.1 **Pathways for converting solar energy**

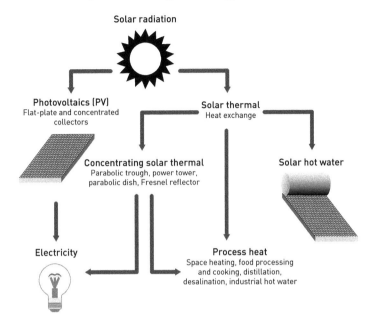

Solar radiation

Photovoltaics (PV)
Flat-plate and concentrated collectors

Solar thermal
Heat exchange

Concentrating solar thermal
Parabolic trough, power tower, parabolic dish, Fresnel reflector

Solar hot water

Electricity

Process heat
Space heating, food processing and cooking, distillation, desalination, industrial hot water

SOURCE ABARE and Geoscience Australia (2010).[9]

is known as Global Horizontal Incidence (GHI) radiation. In contrast, solar concentrators used in CST and concentrated solar photovoltaic (CPV) power stations can only focus direct solar radiation, as illustrated in Figure 2.2, and give negligible output during overcast weather. Direct radiation is measured by the sunshine falling upon a flat plate oriented perpendicular to the direction of the Sun and is known as Direct Normal Incidence (DNI) radiation.

There are four types of concentrating solar collector, shown in Figure 2.2. For CPV, solar cells are placed at the focal point or focal lines of the collectors. Concentrating collectors must be used in solar thermal electric systems in order to obtain a high-enough temperature to generate electricity efficiently. For CST, parabolic troughs collect solar energy in pipes, filled with heat transfer fluid, running along the focal line of the collectors. Troughs are the most mature of the CST technologies and are classified here as early diffusion. The three other types are still at the demonstration stage.

FIGURE 2.2   **Four types of concentrating solar collectors**

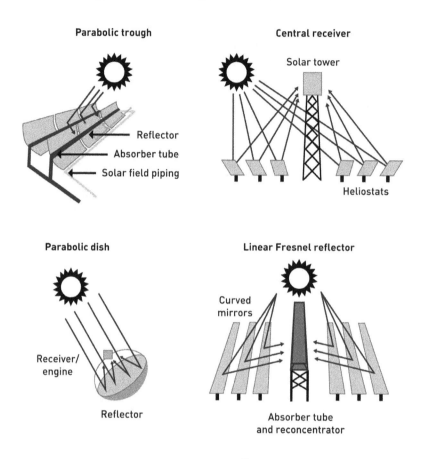

SOURCE IEA-ETSAP and International Renewable Energy Agency.[10]

The solar tower or central receiver and the paraboloidal dish produce the highest concentrations and so are the most efficient of these. They collect solar energy at a 'point' (actually a small surface). Trough and linear Fresnel collectors concentrate sunlight to a lesser degree and so achieve lower temperatures in CST systems, have lower thermal efficiencies and occupy more land per unit of electricity generated. However, experience gained from these systems over the next few years may show that they are less expensive than the more efficient systems and so could have important roles in regions where there is plenty of available land.

## Storage of solar energy

Because insolation fluctuates due to the day–night cycle, changing cloud cover and seasonal variations, it is often useful to have storage as a component of solar energy systems. It is much easier and therefore cheaper to store solar energy as heat than as electricity. Batteries are expensive, although costs are declining slowly. In a passive solar house (Figure 2.3) the solar energy that enters through windows on winter days is stored in the thermal mass of the house in the form of a concrete slab floor or strategically placed solid walls. (In summer, sunshine is excluded from the building by eaves, shutters and blinds.) In solar hot water (Figure 2.4) the solar heated water is stored in an insulated tank. In CST power (Figure 2.5), the high-temperature heat is generally stored in huge tanks of molten salt. Storage by means of concrete or graphite blocks and thermo-chemical processes, such as the thermal decomposition of ammonia into nitrogen and hydrogen, are under development. In place of storage some CST collectors are coupled with gas-fired power stations. Provided there is sufficient land available at the site, CST can also be coupled with coal-fired power stations as a fuel saver – the first of these was built at Liddell Power Station, NSW, Australia in 2008 and expanded in 2012.[11]

FIGURE 2.3  **Passive solar house**

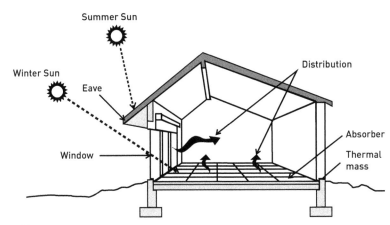

FIGURE 2.4 **Solar hot water: evacuated tube collectors**

SOURCE Ben Elliston.

Because of the high cost of battery storage, almost all solar PV systems on suburban roofs are at present connected to the electricity grid without any local energy storage. The grid itself acts like an energy store. In remote areas solar PV is combined with a diesel or petrol generator and/or batteries in so-called hybrid systems. The retail prices of electricity from the grid are increasing rapidly in some locations while battery prices are decreasing slowly, with the result that within a decade or sooner there may be an economic case for some suburban households and commercial electricity users to install battery storage with PV modules. Initially a small amount of battery storage would be used to supply power during peak demand in the late afternoon and early evening. However, as the prices of batteries continue to fall over the next decade or so, it may become economically viable for suburban households and commercial businesses to install much larger amounts of battery storage to supply 24-hour power and disconnect from the grid. This will cause problems for the current business model of the electricity industry, which is already in trouble, where solar PV and energy efficiency (EE) are reducing the demand for electricity from the grid (see Ch 9, 'New business models for electricity industry').

## Solar PV

A solar PV cell is a solid-state device that converts the energy of light incident upon it directly into electricity by the photovoltaic effect. It has no moving parts. The energy in the sunlight detaches electrons from the atoms of the surface material and they form an electric current. The process is explained in more detail, in non-technical terms, by Andrew Blakers, Foundation Director of the Centre for Sustainable Energy Systems at the Australian National University, in 'What is photovoltaic solar energy?'.[12]

The two principal types of solar PV material are crystalline silicon, which has 85 per cent of the market today, and thin film, which has most of the remaining 15 per cent. There are two principal types of crystalline silicon cells: monocrystalline, made from thin slices cut from a single crystal of silicon, and polycrystalline, cut from a block of silicon crystals. Thin film cells are made by depositing very thin layers of photosensitive materials onto a low-cost backing, such as glass, stainless steel or plastic. Their lower production costs counterbalance this technology's lower efficiency.[13] Other types of solar cell, such as organic solar cells and dye-sensitised solar cells, are still on the early stages of the path towards maturity.

Matching supply and demand by time of day and season of the year is a particular challenge for solar PV without dedicated storage. Peak solar generation varies from over 1 kilowatt per square metre ($kW/m^2$) at noon on a clear day to zero on a moonless night.[14] Peak electricity demand in summer occurs in mid- or late afternoon, partially overlapping peak solar supply. Passing clouds can switch off a solar power station with concentrating collectors and cause considerable fluctuations in the output of one with flat-plate collectors. In the absence of batteries the temporal variations of solar PV can be addressed by increasing the geographic distribution of the solar power stations, introducing demand reduction (see Ch 3) and adding dispatchable power stations to the generation mix – CST with thermal storage in appropriate regions, hydro-electricity with dams and gas turbines – to help smooth the fluctuations.

At the end of 2012, global cumulative capacity of solar PV passed 100 gigawatts (GW).[15] (As indicated in Table 1.4, 1 GW is the rated power of a typical coal-fired or nuclear power station.) Prices of PV modules began to fall steadily from around 2008–09 and then rapidly during 2010–12. Post-2012 they are still falling, although more slowly (see Ch 5, 'Quantified costs and unquantified benefits'). These price reductions are primarily the result of global manufacturing capacity far exceeding demand, with China in particular giving strong incentives to its manufacturers. This has led to an international price war and bankruptcies of several PV manufacturers, including former market leader Suntech in China and Solyndra in the USA. With companies focusing on survival, technological innovation has been placed on the back burner. An industry shake-out and consolidation is under way.

### Concentrated solar thermal (CST) power

In CST, direct sunlight (DNI) is concentrated to heat a fluid that turns water into steam that drives a steam turbine. There are several regions of the world where DNI levels are very high and large solar power stations with concentrating collectors, both CST and CPV, could be installed over vast areas. These include south-west USA and Mexico, the Atacama Desert west of the Andes in South America, much of the Middle East and North Africa, parts of southern Africa, north-west China and west of the Great Dividing Range in Australia. In the future, these regions could become major exporters of solar energy by transmission line and possibly

FIGURE 2.5    **CST power stations**

## (a) Parabolic trough: Part of Andasol 1, 50 MW, in Spain

SOURCE ACS Cobra/ESTELA <www.estelasolar.eu>.

## (b) Power tower (central receiver): Gemasolar, 20 MW

SOURCE Gemasolar solar thermal plant, owned by Torresol Energy © SENER.
NOTE Gemasolar became operational in 2011; 140 m high tower; thermal storage tanks in front of the tower are capable of 15 hours storage following a sunny day.

in the long term by hydrogen tanker ship. In this way solar energy supply could in principle be matched with energy demand over the whole world.

At the end of 2011, global cumulative capacity of CST reached 1.8 GW, driven mainly by Spain. However, in 2012 growth in CST slowed there, as a result of the Global Financial Crisis (GFC) and the unfortunate situation that the Spanish government was responsible for paying for the feed-in tariffs instead of electricity consumers. Another factor was the rapid price reduction of solar PV. Global capacity at the end of 2012 was estimated to be about 2.5 GW.[16] In addition, CST projects, each of at least 20 megawatts (MW), are under construction in 2013 in Australia, China, Egypt, India, Morocco, South Africa, Spain, the United Arab Emirates and the USA.[17]

### Future prospects for solar energy

On a global scale, there is still very large potential for passive solar housing and solar hot water, both fully commercial technologies. Solar heating and cooling of buildings and industrial solar heat with concentrators are just starting to enter the market and their long-term economics are unclear.

PV's low and decreasing prices, huge resource, low environmental and health impacts (see Ch 5), low land use compared with agriculture (see Ch 5) and widely available materials (at least for silicon crystal technologies) gives confidence that, with geographic diversity of collectors, it can make a substantial contribution to future electricity demand.

Although the growth of CST has been slowed temporarily by the success of PV and the impact of the GFC, its capacity is growing from a tiny base in several countries with high DNI that have ridden out the GFC. Therefore, in the longer term, it is likely that its cost will continue to decrease and its advantage of low-cost thermal storage will enable it to play an important complementary role to PV.

# Wind energy

In general the windiest on-shore sites are located near the coast at latitudes higher than about 35 degrees. Exceptions to this general rule include mountain passes, ridges, shores of inland lakes and extensive plains where unobstructed wind can build up speed.

Wind electric power was initiated by Danish school-teacher Poul la Cour at Askov School in 1891. Although his wind turbines were crude designs like traditional 'Dutch' windmills, whose wide blades were pushed around inefficiently by the wind, he passed the electricity through an electrolysis unit and used the hydrogen produced to light the school with gas lamps.[18] Modern wind turbines, with airfoil blades that 'fly' on the wind, evolved steadily in Denmark through the 20th century and set the global standard design of three blades upwind of the tower. Today wind turbines are a reliable technology, wind speeds are predictable to quite a high degree and wind energy is the cheapest of the non-hydro large-scale renewable electricity technologies.

Wind energy is a major industry. In 2008 and 2009 wind power made the largest contribution to new generating capacity in Europe. At the end of 2012, global cumulative installed wind power capacity reached 282.4 GW. Of this total China had 27 per cent, the USA 21 per cent, Germany 11 per cent, Spain 8 per cent and India 6.5 per cent. The growth in total global capacity during 2012 was 18.6 per cent,[19] quite good considering the impact of the GFC, but well below the average growth rate over the previous decade of about 28 per cent per year. According to joint scenarios by Greenpeace International and the Global Wind Energy Council, wind energy could supply up to 12 per cent of global electricity by 2020 and more than 20 per cent by 2030, given effective policies from governments.[20] Wind energy is expected to make particularly large contributions to the official renewable electricity targets of the following countries:

- Denmark: 100 per cent renewable electricity by 2035.
- Scotland: 100 per cent renewable electricity by 2020.

- New Zealand: 90 per cent renewable electricity by 2025.

To achieve such targets in densely populated regions such as northern continental Europe, off-shore sites have to be utilised. Wind speeds are generally higher and less variable off-shore than in nearby on-shore locations. However, both capital costs and maintenance costs of off-shore wind farms are currently about double on-shore costs. At the end of 2012, off-shore wind capacity, included in the above world total, reached about 5.4 GW, of which 1.3 GW was installed in 2012.[21]

In this book off-shore wind turbines with fixed foundations are classified as early diffusion, because there is still much opportunity for gaining experience in construction of components, installation and maintenance, and so bringing down costs. At present almost all off-shore wind turbines are installed on foundations embedded in the sea-floor in waters of depth less than 30 metres and most are located in the North and Baltic Seas. Even with this depth limitation there is very large resource potential. If the costs can be brought down, especially those of the foundations, then the European Wind Energy Association estimates that off-shore wind power could grow to a generating capacity of 150 GW by 2030 and supply 14 per cent of Europe's electricity demand.[22]

Drawing upon the knowledge base of floating oil platforms, a few experimental floating wind turbines have been operating in waters of depths of over 100 metres since 2009. While these are still at an early stage of development, they could, if successful, greatly expand the usable global wind energy resource.

## Bioenergy

Globally, photosynthesis stores solar energy in biomass (organic material) at a rate of about seven times the 2010 rate of global energy use, 532 exajoules (EJ) per year[23] (1 EJ = $10^{18}$ J). So there is a substantial biomass resource worldwide. At present biomass provides about 10 per cent of world total primary energy supply (50 EJ per year). Most of this is energy in villages in poor countries, where twigs, leaves, cow-dung and diminishing supplies of forest timber are burned inefficiently to cook food and to heat living space. This creates much local air pollution and respiratory diseases for the women who burn these fuels, often in unventilated huts.

The principal limitations on the large-scale use of biomass by industrial societies are environmental impacts (see Ch 5, 'Other environmental, health and social impacts'), coupled with the sheer magnitude of potential future demand. However, environmentally unsound practices, such as the production of ethanol from corn (maize) in the USA and the clearing of tropical forests to grow palm oil trees in south-east Asia, need not blind us to the possibilities of using biomass on a limited scale and in environmentally sustainable ways.

A potentially sustainable source of biomass for bioenergy is the use of crop and forestry residues, abundant in countries with large agricultural and forestry industries. After the biomass is converted into useful energy, the ash can be back-loaded to farms on the trucks that collect the biomass, thus returning nutrients to the soil. Another potentially sustainable use of biomass is the planting of energy crops on marginal land unsuitable for agriculture, especially where there are environmental benefits from revegetation and ecological restoration of land that has been previously damaged by clearing, grazing and other activities. Yet another possible example, which is being tried on a pilot scale in Western Australia, is the planting of thin strips of a Eucalyptus species known as Oil Mallee in wheat fields. Although there may be a slight loss in wheat production in the short term, there are long-term benefits where the deep-rooted mallee help mitigate dryland salinity.[24] There may also be short-term benefits to wheat production, if the mallee strips reduce wind speeds at ground level. In Europe woody bioenergy crops such as willow and poplar provide local environmental improvements, including increases in soil organic matter, water quality and habitat for biodiversity.[25] In Australia's Northern Territory it has been proposed to control the prolific mimosa

FIGURE 2.6     **Some pathways for converting biomass into bioenergy**

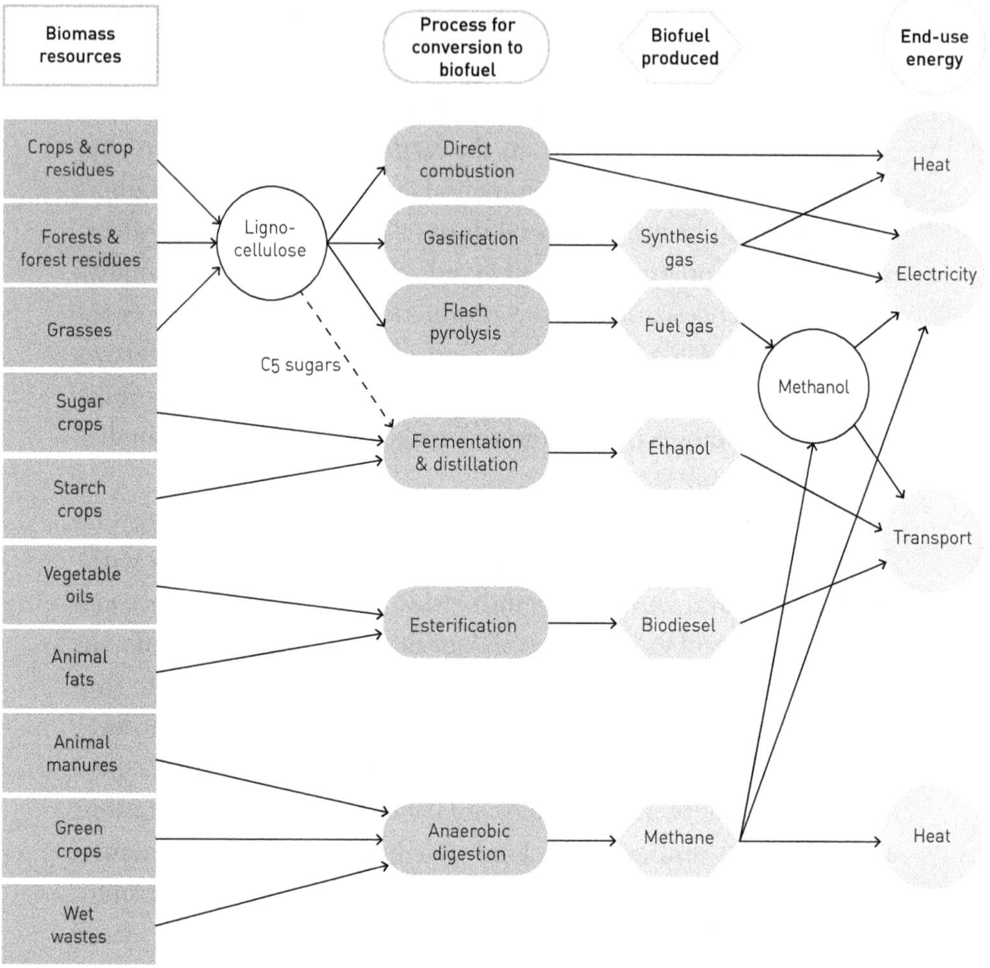

| Biomass resources | Process for conversion to biofuel | Biofuel produced | End-use energy |

Figure contents:
- Crops & crop residues
- Forests & forest residues
- Grasses
- Sugar crops
- Starch crops
- Vegetable oils
- Animal fats
- Animal manures
- Green crops
- Wet wastes

- Ligno-cellulose
- C5 sugars
- Direct combustion
- Gasification
- Flash pyrolysis
- Fermentation & distillation
- Esterification
- Anaerobic digestion

- Synthesis gas
- Fuel gas
- Methanol
- Ethanol
- Biodiesel
- Methane

- Heat
- Electricity
- Transport
- Heat

SOURCE Adapted by the author from Stuckley et al. (2004).[26]

weed by using it as a fuel in power stations.[27]

Biomass can be converted to usable energy in many different ways, some of which are illustrated in Figure 2.6. Solid biomass can be combusted directly to produce heat and electricity (base load and intermediate load), as is common in Scandinavia, Canada, Brazil and parts of the USA. Many coal-fired power stations in the European Union nowadays burn a mixture of coal and biomass, a practice known as co-firing. Domestic heating by burning wood pellets is much more efficient and gaseous fuels are less polluting than traditional wood-burning.

Biomass can be converted to liquid fuels methanol, ethanol, butanol and biodiesel. Ethanol and butanol are produced from sugars and starches by fermentation and distillation, a commercial process, but inefficient in terms of land use. Research, development and demonstration are proceeding on extracting sugars from ligno-cellulose, the non-edible fibrous part of plants, which would convert a much greater proportion of the plant into biofuel and hence yield a much higher efficiency of land use. Other R&D in progress includes the production of liquid fuels from algae and by genetically engineering microbes to convert sugars into specific fuels. Biodiesel is made commercially from vegetable oils by the simple chemical process of esterification, but it too is limited by land-use constraints. With quite high energy densities (but ethanol and methanol are significantly below that of petrol – see Table 7.2) and ease of handling, these liquid biofuels can be combusted conveniently in gas turbines and gas engines for stationary energy and motor vehicles. Biodiesel can also fuel jet aircraft. Biofuels are likely to be more convenient forms of energy storage and transportation than hydrogen in its gaseous, liquid or metal hydride forms.

Biomass can also be converted into gaseous fuels. Wet biomass (green crops, animal manures and other moist wastes) can be converted to biogas, which is mostly methane ($CH_4$), by anaerobic digestion; after upgrading to a quality similar to fossil natural gas it can be fed into gas pipelines and used for heating. A different renewable gas, called synthesis gas or syngas, can be produced by heating dry biomass in limited oxygen. The product is a mixture of mainly carbon monoxide (CO) and hydrogen ($H_2$), as well as some $CO_2$, water and other hydrocarbons. Syngas can be combusted in gas turbines or converted to methane and fed into gas pipelines, or converted to methanol.

Like fossil fuels, biomass can also be used as the raw material for making plastics, fertilisers and other chemicals. Further down the track is the possibility of biorefineries that produce a diversity of products – electricity, heat, chemicals and liquid and gaseous biofuels.

While debates about the sustainability of 'first generation' biofuels continue, there are potentially large future markets for advanced biofuels from agricultural and forestry wastes, and from crops grown on marginal or otherwise-unproductive lands. Already there are expanding international commodity markets for wood pellets and bio-oil for heating, greater use of biogas in a variety of applications and greater use of biomass in heat supply. Table 2.2 summarises the technological maturity of various bio-energy technologies.

TABLE 2.2　**Technological maturity and potential of various bioenergy technologies**

| Technology | Status | Potential | Environmental impacts |
|---|---|---|---|
| Combustion of solid biomass for electricity and heat | Commercial | Medium | Low from sustainable sources |
| Biogas (methane) by anaerobic digestion of wet organic wastes | Commercial | Low | Low |
| Landfill gas | Commercial | Low | Low |
| Ethanol from sugars and starches, 1st generation | Early diffusion | Low from sustainable sources; medium otherwise | Low from sustainable sources; high otherwise |
| Methanol | Early diffusion | Medium to large | Low from sustainable sources |
| Gasified biomass for combustion in gas turbine or gas engine | Early diffusion | Medium | Low from sustainable sources |
| Ethanol, 2nd generation, from ligno-cellulose | R&D and early demonstration | Medium to large | Low–medium |
| Biodiesel from algae | R&D | Medium | Low |
| Microbial production of hydrogen | Research | Large if economic | Low |

SOURCE The author's review of the literature.

# Hydro-electricity and marine energy

Hydro-electricity currently supplies about 16 per cent of global electricity so it is a commercial technology of long standing. Major undeveloped resources still exist in China, South America (Amazon Basin) and central Africa (Congo River), although almost all could have severe environmental and social impacts if developed.

The principal type of hydro-power station has a reservoir and so can be operated to meet peaks, to ramp up power supply very quickly when another power station breaks down and to compensate quickly for fluctuations in RE sources without storage. Thus it is an ideal complement to wind and solar PV. If the reservoir is big enough, it can provide seasonal storage of energy.

Another type of hydro-electric system is pumped storage. This has two reservoirs with a large height difference. During periods of low demand when electricity is cheap, water is pumped from the low to the high reservoir. When there is a peak in demand or a failure in supply, electricity is more expensive and water is released from the higher reservoir, generating peak power on its descent to the lower reservoir. If the pumped storage system is only required for compensating for fluctuations in supply and demand over timescales of hours to days, the storages can be quite small, making many more potential sites available than for a large dam.

Other types of hydro station have no storage and comprise run-of-river systems. Their turbines are simply immersed in a river or in a channel constructed to take some flow from a river. They are often small systems generating a few kilowatts.

Marine power stations are of several types. Their respective maturity statuses, potentials and environmental impacts are summarised in Table 2.3. Only conventional tidal power is fully commercial; however, it is restricted to estuaries with very high tides, such as in Brittany (France), where La Rance station of 240 MW has been operating since 1966; the west coast of South Korea, where the Sihwa Lake station of 254 MW was opened in 2011; the Severn Estuary in the UK, where a mega-project has been proposed to

TABLE 2.3    Types of marine power stations

| Technology | Status | Potential | Environmental impacts |
| --- | --- | --- | --- |
| Tidal reservoir | Commercial | Geographically limited | High |
| Ocean and tidal current | Demonstration | Geographically limited | Low |
| Wave (several types) | Demonstration | Medium to large | Low |
| Ocean thermal energy conversion | R&D | Geographically limited | Low |

SOURCE The author's review of the literature.

take advantage of the huge tidal range (up to 14 metres); the Bay of Fundy in eastern Canada; the Gulf of Cambay (or Khambhat) on the west coast of India; and the north-west coast of Australia which is very remote from population centres.

Ocean and tidal current systems have underwater turbines placed in fast-moving currents, with no dams. Fast currents are sometimes formed in estuaries, straits and around headlands. Water speed is critical, because, as for wind power, water power is proportional to the cube of the speed. Although water speeds are generally much lower than wind speeds, the density of water is much higher and power is proportional to fluid density. Scotland has potential and is a world leader in trying out this technology, with several demonstration plants operating. Another possible source is the mighty Kuroshio Current that runs up the east coast of Taiwan to Japan. It has huge potential and an initial pilot plant of 30 MW is planned in Taiwan.

Wave power is less geographically limited than conventional tidal and ocean/tidal current. The greatest wave power densities occur off northwestern Europe, western Canada, Alaska, southern South America and southern Australia. At present many types of converter are being demonstrated around the world, for example, one described as a nodding floating 'duck'; another resembling a string of floating sausages; another mimicking a field of plants attached to the ocean bottom and oscillating back and forth; and yet another in the form of a hydro-electric reservoir on a coastal cliff into which waves are channelled by means of underwater concrete blocks near the base of the cliff. It is not yet clear which of these technologies will stand up best to the rigorous marine conditions and which are the most cost-effective. In most cases the movement induced by waves is converted into electricity.

Ocean thermal energy conversion (OTEC) would use a heat pump (see Box 2.2) running off the small temperature difference between warm surface water in a tropical region and deep cold water. Since very few regions satisfy these conditions, the potential appears to be very limited at present. Hawaii is one of the few places where this may be feasible and there are plans for demonstration plants there. Heat from the warm surface water is used to vaporise a fluid such as ammonia, which turns a turbine to drive a generator to produce electricity. Because OTEC converts a low-quality (low-temperature) energy source into a high-quality source (electricity), its thermal efficiency is very low, typically less than 5 per cent, and this could make the economics difficult. However, power production would be continuous, without large fluctuations.

Since all the marine sources are either geographically limited or still at the demonstration stage, they do not play a significant role in the future global energy scenarios discussed in Chapter 3. However, in particular regions they could turn out to be important. The next decade should clarify their respective potentials.

---

### Box 2.2    The heat pump

Here is a simplified description, starting with a well-known heat pump, the refrigerator. It circulates a fluid called a coolant or refrigerant, which boils at low temperature. This property allows it to collect heat from inside the compartment and dump it outside from the coils at the back of the fridge and then return the coolant to the compartment where it collects more heat. This is how it does it (see Figure 2.7).

FIGURE 2.7 **Operation of a refrigerator**

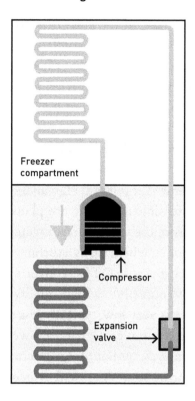

SOURCE Science Treasure Trove.
NOTE Fluid in the bottom half of the diagram is in gaseous phase; fluid in the top half is liquid phase. The shaded area comprising the bottom half of the diagram is outside the compartment.

The coolant is piped into the fridge compartment in liquid form under low pressure. It flows through the coils of a heat exchanger, absorbing heat from inside the fridge compartment. This heat vaporises the coolant without changing its temperature, in the same way that boiling water remains at a fixed temperature. The coolant is then compressed by a pump called the compressor. This heats it up. It then passes through coils of pipes on the outside of the fridge compartment, either at the back or underneath. Because it is hotter than the room, it releases heat into the room. As it cools it changes back into a liquid, still under pressure from the compressor. It is then pumped back into the compartment through an expansion valve, a tiny hole. On the other side of the hole is low pressure, so the liquid boils and vapourises, picking up heat, and so the cycle continues. External energy has to be supplied in the form of electricity to run the compressor, but this energy flow is much less than that pumped out of the fridge.

An air conditioner is also a heat pump, pumping heat from inside a room to the outside the building. In reverse-cycle operation it pumps heat from outside to inside.

## Geothermal energy

About 6–7 metres below the Earth's surface the temperature is approximately constant over days and seasons. So the ground can be used as a stable temperature reservoir to centrally heat and cool buildings by means of a heat pump. Pipes are buried near the building, so that the ground at 6–7 metres can be used as a heat source in winter and heat sink in summer. Air is the medium that carries the heat or 'coolth' through the pipes. Such systems are known as *ground-source heat pumps* or *geothermal heat pumps*, although the heat does not come from deep in the Earth, but from the Sun. These are commercial systems and are generally more economical for large buildings, rather than small.

An entirely different kind of geothermal system is the use of the Earth's heat at high temperatures to generate electricity. Conventional geothermal power exists in volcanic regions where steam and/or hot water come to the surface of the Earth. Steam drives a turbine to generate electricity. Conventional geothermal potential is a commercial technology; however, it is limited to volcanic regions. It supplies quite large shares of total electricity demand in Iceland (25 per cent), El Salvador (22 per cent), Kenya (17 per cent), the Philippines (17 per cent) and Costa Rica (13 per cent). In absolute terms, the USA generated the most geothermal electricity in 2009: 16.6 terawatt-hours (TWh) from an installed capacity of

3 GW. About 11 GW is installed worldwide.[28] There is considerable untapped potential in Indonesia and Japan.

A much greater potential could be obtained by extracting the heat in hot granite rocks several kilometres below the surface. It involves drilling two deep wells, cracking (actually 'fracking') the rock between their bottoms, pumping water down one well to extract the heat and bring it to the surface as steam via the other well to generate electricity on the surface. This process is known as *enhanced* or *engineered* or *hot rock geothermal power*. Most continents have extensive underground regions with granite rocks at temperatures of 150–250°C. Most of the stored heat comes from the decay of traces of radioactive elements in the rock. If there is an insulating layer above the granite, the temperature gradually increases over many millions of years.

At present there are only a few small-scale (1–3 MW electrical)[29] demonstration plants worldwide: in France, Germany and Australia. This technology is likely to develop slowly, because the costs of drilling are high and the development process is risky in terms of exploration and establishing the water circuit. Although geothermal power is not, strictly speaking, renewable, it is generally classified as such because the heat extraction is small compared with the huge size of the heat reservoirs.

## Storage of renewable electricity

The storage of solar energy as heat is the simplest and cheapest means of storage, as outlined in 'Solar energy', above. Liquid and gaseous fuels, which store energy in chemical bonds, are discussed in the context of transport in Chapter 7, 'Improved vehicles and fuels'. In the present section we focus on storing electricity, a high-grade form of energy. In an RE future, electricity will play an even greater role than it does now, because it is currently the most feasible way of providing urban transport and high-temperature heat from renewable sources. However, before discussing how to store it, and how to avoid the expense of storing it where possible, we describe the fluctuations in RE sources that indicate the need for more storage.

Most RE sources come directly or indirectly from the Sun. While the variations in the brightness of the Sun are tiny, the insolation received at a point on the Earth's surface can vary between about 1 kW/m² at noon in some regions and zero overnight. Due to clouds it can also vary suddenly in daytime at a specific location and, due to the tilt of the Earth's axis of rotation and other factors such as monsoonal rains, it varies slowly by season. Because wind energy is an indirect form of solar energy, it varies less abruptly than insolation at a specific location on timescales of minutes and hours. However, it is strongly influenced by changing weather patterns, which have typical timescales of 5–7 days. It may also vary by season and year. Because waves, except tsunamis, are generated by winds, they vary more slowly than wind. Solar heat stored within the top several metres of the ground varies on a seasonal timescale, which makes it useful for heating and cooling buildings by means of an electric heat pump. Geothermal heat in hot rocks at depths of several kilometres (which are heated mainly by radioactive decay, not stored solar energy) only varies at a specific location on decadal timescales as a result of the source being gradually depleted by generating geothermal electricity. Some types of biomass are only available seasonally and may be expensive to store for the remainder of the year.

These variations entail that either storage or flexible back-up or a combination of both must play a greater role in renewable electricity systems than in conventional systems. Because storing electricity in batteries is currently expensive, it makes sense to give priority to reducing the need for batteries wherever possible.

## Reducing the need for batteries

The first way to reduce the need for batteries is to distribute the RE collectors geographically and feed the electricity generated into the grid. This is just rationalising a necessity, because RE sources are more diffuse than fossil or nuclear sources. The spatial dispersion of large solar and wind power stations smooths out the power fluctuations experienced at a single site on timescales of minutes and hours.[30]

The second means of reducing the need for expensive electrical storage is to further improve weather forecasts and models. Great improvements have already been made over recent decades by the use of satellite data and sophisticated computer models.[31] As a result, wind speeds can be predicted quite accurately from hour to hour and reasonably well 24 hours ahead. The improvement in prediction reduces the need for base-load power stations and makes it much easier for fast, flexible, peak-load power stations to fill the troughs that occur from time to time in solar and wind power supply.

Indeed, flexible power stations, in the form of gas turbines or diesel generators, together with a reliable source of fuel, can be considered to be the cheapest form of storage for handling fluctuations in electricity supply. Flexible generators can be started up rapidly, have low capital costs and, provided they are only operated occasionally when really needed, have low running costs, even when the fuel is quite expensive. Thus their role is that of reliability insurance with a low premium. Some computer simulations of large-scale electricity systems with high penetrations of RE find a vital role for such flexible generators in compensating for the fluctuations in wind and solar power (see Ch 3).

To store effectively high-quality energy forms, such as those required for electrical and mechanical energy, requires low loss while in the store (called 'standing losses') and low loss in transfer to and from the store (called 'cycle efficiency'). The latter is generally measured on a scale of 0 to 1, where cycle efficiency equal to zero means that all the energy transferred to store is lost, while cycle efficiency equal to 1 is the ideal but unachievable case when no energy is lost in transfer. (Cycle efficiencies are less important when storing low-quality energy, that is, heat at low temperatures.) Storing renewable electricity by converting it into heat is not a

TABLE 2.4 **Properties of different forms of storage of electrical and mechanical energy**

| Storage form | Maturity | Storage period | Energy density | | Cycle efficiency |
|---|---|---|---|---|---|
| | | | kJ/kg | MJ/m³ | |
| **Mechanical** | | | | | |
| Storage dam and pumped hydro, 100 m head | Commercial | Large dams provide long-term storage (seasonal or longer); small dams for short-term storage | Low: 1 | Low: 1 | 0.65–0.8 |
| Flywheel – steel | Early diffusion | Several minutes | 30–120 | 240–950 | 0.9 |
| Flywheel – advanced carbon fibre | Demonstration | Several minutes | >200 | >100 | 0.95 |
| Compressed air | Early diffusion | Long term | – | ~15 | 0.4–0.5 |
| **Electrochemical** | | | | | |
| Lead–acid battery – rechargeable | Commercial | Low standing loss; short cycle life | 40–140 | 100–900 | 0.75 |
| UltraBattery – rechargeable lead–acid with capacitors | Demonstration | Low standing loss; medium cycle life | – | – | 0.7–0.8 |
| Nickel–cadmium battery – rechargeable | Commercial | Highly reliable; low standing loss | 350 | 350 | 0.6 |
| Nickel–metal hydride battery – rechargeable | Commercial | – | 250 | 1080 | 0.5 |
| Lithium-ion – rechargeable | Commercial | – | 720 | 1400 | 0.7 |
| Zinc bromine battery – rechargeable – energy stored in electrolyte | Demonstration | No standing loss; high cycle life | 250 | <216 | 0.65–0.8 |
| Zinc–air – rechargeable | Demonstration + new research | – | >1350 | – | 0.45 |
| Vanadium redox battery – rechargeable – energy stored in electrolyte | Early diffusion | No standing loss | 72–108 | 90–144 | 0.8 |

SOURCE Mainly Sørensen (2011a).[32]
NOTE Different sources give very different estimates of energy densities.

good idea, because the conversion of the heat back to electricity is generally inefficient. Compressed air storage is also inefficient, as shown in Table 2.4, because much energy is lost as heat resulting from the compression. It is assumed that the hydrogen ($H_2$) in the table is produced by electrolysis, also an inefficient process. However, alternative processes are under development.

Hydro-electric storages store a high-grade form of energy (potential energy of water due to its height above the power station) ready for rapid conversion into another form of high-grade energy (electricity) with high efficiency (see 'Hydro-electricity and marine energy', above).

### Batteries

Conventional lead–acid batteries have low energy density, medium cycle efficiency, significant standing losses and short lifetimes. An improved version is the UltraBattery, developed by the Commonwealth Scientific and Industrial Research Organisation (CSIRO) in Australia and being demonstrated in several locations. It contains capacitors that enable charging and discharging to occur at high power and so has a longer lifetime.

In the so-called 'flow batteries', vanadium redox and zinc bromine, energy is stored in the electrolyte and so the quantity of energy stored can be increased easily by increasing the size of the tank and adding more electrolyte. Hence the cost per kilowatt-hour declines as the storage capacity increases. Since they have high cycle efficiency and low energy densities, they are suitable for stationary storage of electricity on utility scale.

As the market expands for batteries, they are gradually becoming less expensive. Meanwhile prices of electricity from the grid are increasing in many locations, driven primarily by increasing demand of air conditioning and hence upgrades in the distribution system and more peak-load generators. Although a few companies are now offering packages of solar PV and batteries, the economics of adding batteries are still poor for residential solar in the suburbs.[33] However, if batteries were valued for avoiding capacity upgrades of the distribution network, many more people would buy batteries.[34]

## Conclusion

The Sun could supply, directly and indirectly, many times the current energy demand of industrial society, without significant impact on food production. Technologies exist for converting these safe natural flows of energy into usable energy for an industrial (or post-industrial) society. Hydro-electricity, wind, passive solar housing, solar hot water, solar PV modules, some forms of bio-energy and ground-source heat pumps, together with a wide range of EE technologies and measures, are commercially available on a large scale. One type of CST system with thermal storage and several types of bioenergy system are commercially available on a limited scale. Taking all these together, there is sufficient basis for undertaking a rapid transformation of the energy system to one based predominantly on sustainable energy, that is, the efficient use of RE. The energy supply side of the Great Transition is technolog-ically feasible now.

While this transition is occurring, other RE sources and sup-porting technologies will move from the demonstration and early diffusion stages to large-scale commercial maturity. These are likely to include several wave power technologies, enhanced/engi-neered geothermal power, additional forms of bioenergy that are environmentally sound, new and improved types of energy stor-age, and possibly solar space heating and cooling.

This technological diversity will be needed if the whole of human civilisation is to move to 100 per cent RE, because RE resources, like fossil fuels and uranium, are not distributed uni-formly over the globe. Different regions will need different mixes of RE. Some regions will have to import RE, while others will export.[35]

More detailed accounts of the various energy supply and energy using technologies are given in Chapters 4–7. But first, Chapter 3 explores scenarios for sustainable energy futures, drawing upon those technologies introduced in this chapter that are commercially available, either fully or in limited production.

# 3: SUSTAINABLE ENERGY SCENARIOS

Your task is not to foresee the future, but to enable it.

Antoine de Saint-Exupéry[1]

What if your nation or local community undertook a transition to 100 per cent sustainable energy, that is, the efficient use of renewable energy (RE), by 2050? What energy conversion and energy use technologies would be commercially available? What would the transition cost and how would it be financed? What government policies would you need to drive the transition, in terms of economic incentives, finance, regulations and standards, provision of infrastructure, research and development, education and training, and setting up new institutions? If you chose a set of answers to these questions and explored the consequences systematically, you would be performing a scenario study.

More generally a scenario is a hypothetical sequence of future events. An honest, useful scenario clearly sets out the assumptions for that sequence, in this case for making a transition to a sustainable energy future, and investigates the effects of varying the assumptions. Thus scenarios are not predictions of the future, but rather 'what if' exercises. They are one of the first steps in the necessary enabling process mentioned by Antoine Saint-Exupéry. They allow us to explore the possible consequences of following certain pathways into the future. They are valuable for investigating the sensitivity of the outcomes to different assumptions. If the variables can be quantified, the scenario can be tracked with a computer model. One of the comparison or reference scenarios can be chosen to be 'business-as-usual' (BAU), that is, one with no new policies.

Scenario studies have been valuable in refuting several myths that have been disseminated against RE, namely:

- *Myth 1*: RE is too dispersed, too low density in terms of energy supply per square kilometre, to supply our cities and industries.
- *Myth 2*: RE is 'intermittent' and so large penetrations of RE into the grid cannot provide electricity with the same reliability as fossil or nuclear power.
- *Myth 3*: RE is too expensive to use on a large scale.

First let's look at the knowledge being gained from scenario studies, including hour-by-hour computer simulations, and then see how this knowledge busts the above three myths.

## Overview of scenario studies

Many different types of scenario study contribute to our understanding of low greenhouse gas (GHG), sustainable energy futures. They address individual and regional groups of provinces/states, individual and regional groups of countries, and the whole world. Some focus on the electricity industry, others examine the whole energy sector, and a few others study the whole socio-economy. Some investigate the feasibility of further developing and deploying single technologies, while others research the integration of a portfolio of new and improved technologies. Scenario studies are not restricted to being technical. Part of the research may be on the policies needed from governments and on the strategies that communities can use to push reluctant governments to take effective action.

Scenarios can be classified into forecasting models and backcasting models. Forecasting scenario models start from the present situation and, under various assumptions, make incremental changes at the margin in a desired direction. They may lack a vision of a very different long-term future. Instead they are often strongly influenced in their choices of what is possible by past experience, for instance the historical transitions from the dominance of bio-mass (firewood and other organic fuels) to that of coal and then to oil. Many forecasting scenarios are tied to existing behaviours and institutions – such as consumption patterns, investment patterns, business models, taxation system and electricity market rules – and tend to view radical change as impractical.

In contrast, back-casting scenarios start with a vision of the future that may be very different from the present. Such a vision is essential for mitigating climate change. Visions of an ecologi-cally sustainable future start with assumptions, either explicit or implicit, about the kind of social and economic systems that could exist in that future. For instance, a chosen sustainable future may have no growth in population, a stabilised or reduced through-put of energy and materials, greatly improved technologies, new business models, new rules for electricity markets and indeed for markets in general, innovative methods of finance, new govern-ment structures and new governance processes. Or, if it is a more cautious scenario, it may be based on existing institutions and clean technologies that are commercially available now, and then simply project the prices of these technologies into the future as the size of their markets increase. The back-casting process then involves exploring possible transition pathways from the present to the future vision or visions and the policies needed to follow them.

Back-casting scenarios differ in the degree of detail given to the transitional period between now and the year of the future vision. Some have no intermediate steps while others may proceed in one-year or five-year time-steps, so that the scenarios can track the gradual replacement of technologies such as individual fossil-fuelled power stations and fleets of petroleum-fuelled motor vehicles. In practice, scenarios may involve a mixture of forecasting and back-casting. But without credible visions of a sustain-able future and strategies to achieve them, it will be impossible to avoid devastating, irreversible changes to Earth's climate.

Table 3.1 summarises the assumptions and principal results of a sample of scenario studies published before mid-2013. It is a rapidly expanding field. I have selected these studies because they address integrated systems of RE supply, either for electricity generation or for the whole energy system. Some include demand reduction; many calculate the achievable GHG reductions. Most give a snapshot of a possible sustainable energy system or economy at a fixed time in the future, such as 2030 or 2050, as the first step in a back-casting scenario, while others also explore the dynamic transition process over decades.

Included in Table 3.1 is Chapter 10 of the Intergovernmental Panel on Climate Change (IPCC) *Special Report on Renewable Energy Sources and Climate Mitigation*, published in 2011. Although it briefly reviews 164 RE scenarios, most of these focus on individual RE technologies and so are less relevant to our discussion, which is concerned with replacing all or most fossil fuels with a portfolio of RE and energy efficiency (EE).

Let's now examine several scenario studies in more detail, starting with the economy-wide reduction in GHG emissions, then look at energy and then electricity. After addressing technological and economic aspects we will go on to do some myth-busting. We'll talk about policy recommendations arising from these scenario studies in Chapters 8 and 9.

## GHG reduction scenarios: wide scope

The first study sets out a GHG reduction scenario that spans all sectors of a national economy that are responsible for emitting significant quantities of GHG.

### Zero Carbon Britain

Ultimately scenarios for climate mitigation must be understood in the broad context of the whole socio-economy. The 2010 study *Zero*

TABLE 3.1 **Sustainable energy/climate futures scenario studies**

| Geographic region studied | Year published | Authors and references | Sectors studied | Type of study | Principal results |
|---|---|---|---|---|---|
| **Whole world** | | | | | |
| | 2006 | Stern | Whole economy | Economic analysis at a macro level | Climate change is the greatest and widest-ranging market failure ever seen. Costs of doing nothing are much greater than costs of mitigation. |
| | 2000 | Sørensen and Meibom | Energy | Geographic Information System to match supply and demand | Spatial match feasible between energy demand and 100% RE, given some trade in RE. |
| | 2010 | International Energy Agency (IEA) | Energy | Technology forecasts based on expert knowledge | Blue Map Scenario halves worldwide energy-related $CO_2$ emissions by 2050 with a mix of energy efficiency, carbon capture and storage (CCS), RE, nuclear and fuel switching in that order. |
| | 2011 | IPCC | Energy | Review of literature; mostly technology forecasts | See 'Overview of scenario studies', above. |
| | 2011 | Jacobson and Delucchi | Energy | Resources, technologies, materials use, costs, policies | Assesses energy system characteristics, resources, land and material requirements, options for addressing variability, economics. Finds 100% renewable electricity (RElec) from wind, solar and hydro by 2050 is feasible and affordable. |
| | 2011 | WBGU | Energy/ climate | Strong focus on international strategy and policy | A new global social contract is needed and achievable for a low-carbon and sustainable global socio-economic system. This includes capping the usual options for economic growth in the rich countries. |
| | 2011 | WWF, Ecofys and OMA | Energy | Ecofys scenario; technologies, partial economic analysis | Given that global energy demand stabilises by 2020 and then contracts slightly, 100% of energy supply can be renewable by 2050. |
| | 2012 | GEA, Ch 17 | Energy | Combination of back-casting and forecasting. Technologies, policies and investment costs | It is technically possible to achieve improved energy access, air quality, and energy security simultaneously while avoiding dangerous climate change. See 'GHG reduction scenarios: wide scope', above. |

| Geographic region studied | Year published | Authors and references | Sectors studied | Type of study | Principal results |
|---|---|---|---|---|---|
| | 2012 | Greenpeace, GWEC and EREC | Energy | Forecasting and back-casting scenario to 2050, considering technologies, economics and policies | By 2050 almost all global electricity and heat, and nearly half global transport energy, could be supplied by RE. Global $CO_2$ emissions could be reduced to 15% of the 1990 level. |
| **Europe** | | | | | |
| | 2010 | ECF | Electricity | Roadmap; back-casting | At least 80% reduction in GHG emissions is possible for Europe by 2050, provided electricity sector becomes zero carbon and EE program continues. |
| | 2011 | Heide et al. | Electricity | Hourly simulation over 8 years | 100% RElec from wind and solar only. Explores the synergy between storage and balancing requirements. |
| | 2012 | Rasmussen et al. | Electricity | Hourly simulation over 8 years | 100% RElec from wind and solar only. Explores the synergy between storage and balancing requirements. |
| **Individual countries and neighbours** | | | | | |
| **Australia** | 2012–13 | Elliston et al. | Electricity | Hourly simulations over 1 year | Reliability achieved by different mixes of commercially available wind, photovoltaic (PV), concentrated solar thermal (CST) with thermal storage, existing hydro and biofuelled gas turbines. Economics of RElec mix compared with efficient fossil scenario. |
| | 2013 | AEMO | Electricity | Hourly simulations and probabilistic modelling | Costs evaluated subject to reliability criterion. No comparison scenario with fossil fuels. |
| **Denmark** | 2009 | Lund and Mathiesen | Energy | Expert inputs; analysis | 100% RE is feasible by 2050. |
| **Germany and neighbours** | 2011 | SRU | Electricity | Simulations; scenarios | SRU (German Advisory Council on the Environment) shows 100% RElec technically feasible, reliable and affordable for Germany, but less expensive for regional networks such as Germany–Scandinavia or Europe–North Africa. |

| Geographic region studied | Year published | Authors and references | Sectors studied | Type of study | Principal results |
|---|---|---|---|---|---|
| Ireland | 2011 | Connelly et al. | Energy | Deterministic input–output model integrating all energy; 4 scenarios | Subject to the assumptions, several 100% RE scenarios are technically feasible. |
| Japan | 2003 | Lehmann | Energy | Simulations of 6 scenarios over 1 year with 15 min. time-step. | 100% RE technically feasible, assuming storage from hydrogen produced locally or imported. |
| New Zealand | 2010 | Mason et al. | Electricity | Simulation | 100% RElec achievable in a system that already has mostly hydro. |
| Northern Europe | 2008 | Sørensen | Energy and hydrogen ($H_2$) storage | Simulations over 1 year with 6 hour time-step; with and without trading between countries | 100% RE technically feasible with seasonal hydro storage. Transport could be achieved either with biofuels, or, assuming technical advances, with hydrogen. Large scope for energy trade between countries. |
| Portugal | 2011 | Krajačić et al. | Electricity | Hourly simulations of 3 scenarios spanning 1 year | 100% RElec is technically feasible in systems with hydro storage. |
| UK | 2010 | Kemp and Wexler | Integrated energy and land use; technologies and policies | Radical scenario and policies | An integrated plan for zero carbon Britain by 2030 achieving 90% emissions reduction and 10% carbon capture in agriculture and forestry, by reduced demand and RE supply. See 'GHG reduction scenarios: wide scope', above. |
| USA | 2012 | National Renewable Energy Laboratory (NREL) | Electricity | Hourly simulation | Renewable electricity generation from commercially available technologies, with a more flexible electric system, could supply at least 80% of total US electricity generation in 2050 while meeting electricity demand on an hourly basis in every region of the USA. |

NOTES 'Energy' comprises electricity, transport and non-electrical heat. Scenarios are back-casting, unless labelled forecasting.
KEY TO REFERENCES Stern (2006); Sørensen and Meibom (2000); IEA (2010); IPCC (2011); Jacobson and Delucchi (2011); WBGU (2011); WWF, Ecofys and OMA (2011); GEA (2012); Greenpeace, GWEC and EREC (2012); ECF (2010); Heide et al. (2011); Rasmussen et al. (2012); Elliston et al. (2012 & 2013); AEMO (2013); Lund & Mathiesen (2009); SRU (2011); Connelly et al. (2011); Lehmann (2003); Mason et al. (2010); Sørensen (2008); Krajačić et al. (2011); Kemp and Wexler (2010); NREL (2012).[2]

*Carbon Britain 2030* (ZCB2030), from the Centre for Alternative Technology in Wales, is one of the few recent national studies that tackles this mammoth task.[3] Based on a mix of technologies that are either commercially available or at the demonstration stage, it describes in detail a feasible vision of Britain in 2030 with no net GHG emissions. It addresses both demand and supply. On the demand side, the transformed socio-economy uses energy more efficiently, especially in buildings; has a reduced demand for travel by private car; has no domestic air travel; and consumes less meat from cows and sheep. Nevertheless, as Rob Hopkins, founder of the Transition Towns movement, points out in the Foreword, 'ZCB2030 ... is not a hair shirt, survivalist rejection of modernity, rather it is the logical, achievable next step forward ...'. There is enough food and employment for all.

On the supply side, the scenario's electrical and heating energy are generated entirely by renewable sources. The principal sources of renewable electricity, ranked from the largest contribution downwards, are off-shore and on-shore wind, bioenergy from various sources, wave and ocean current, and tidal with dam. Most of the heat requirement comes from electricity and most of this is used to drive heat pumps. The remainder of heat comes from biomass and solar hot water. Of course, in different parts of the world the energy mixes for electricity and heat may be very different from those in Zero Carbon Britain.

ZCB2030 has a much greater use of public transport, walking and cycling than at present, while the reduced use of private cars is fuelled mostly by renewable electricity. The well-integrated scenario uses the land saved from reduced livestock to grow biomass for both energy and carbon sequestration. The report focuses on technologies and systems, discusses policy options, but only touches on economic aspects.

## Global Energy Assessment

The Global Energy Assessment (GEA), published in 2012, 'aims to facilitate equitable and sustainable energy services for all, in particular the two billion people who currently lack access to clean, modern energy'. It is a huge body of work, with 34 expert lead authors coordinated by the International Institute for Applied Systems Analysis (IIASA) and supported by government and non-government organisations, several United Nations organisations, the World Bank and industry.[4] The book version weighs over 5 kilograms!

GEA developed and explored '... 60 alternative pathways of energy transformations toward a sustainable future that simultaneously satisfy all its normative social and environmental goals of continued economic development, universal access to modern energy carriers, climate and environment protection, improved human health, and higher energy security'. These are quantitative scenarios, expressed through detailed computer models. The scenarios all assume implicitly that they have political commitment and the necessary finance, and that improvements in technologies are disseminated over the whole world. The scenarios have different mixtures of demand-side and supply-side technologies and measures. GEA classified these scenarios into three groups:

- GEA-Efficiency, in which reductions in the demand for energy services and demand-side EE measures play a major role;
- GEA-Supply, in which demand is higher and emphasis is on new supply technologies, such as carbon capture and storage (CCS), to satisfy the additional demand;
- GEA-Mix, which has greater regional diversity in energy mixes, with some regions similar to GEA-Supply and others to GEA-Efficiency.

All 60 scenarios were designed to meet GEA's climate mitigation requirement of limiting global warming to less than 2°C above the pre-industrial level with a probability of at least 50 per cent.

However, as GEA recognises, this is unlikely to be a safe level, so it is doubtful that the GEA scenarios are truly sustainable. Of the 60 scenarios, 41 fulfilled all the normative goals set for the study, showing that there are many ways of providing a technically feasible, affordable, environmentally more sustainable and socially more equitable energy future compatible with available resources. The study found that scenarios that emphasise demand reduction were more likely to achieve these multiple goals, including specifically reduced local and indoor air pollution resulting in significant health gains, reduced traffic congestion, reduced poverty, productivity gains, increased comfort and well-being, potential energy security improvements, new business opportunities and sometimes enhanced employment opportunities. While nuclear power contributed to some scenarios, it was found to be unnecessary for meeting the climate target, even in the high-demand scenarios. On the other hand, GEA found that the high-demand scenarios, including the present trajectory, need contributions from CCS, which is just on the threshold of the demonstration stage.

GEA estimated that the transformation of the energy system would require dedicated efforts to increase global energy-related investments to between US$1.7 trillion and US$2.2 trillion annually, compared with about US$1.3 trillion in annual investment today. The GEA-Efficiency scenarios required the lowest investment of the range and the GEA-Supply the highest. Compared to the BAU scenario that lacks policies to achieve the GEA objectives, the projected loss to consumption by 2050 is relatively small, ranging from 0.6 per cent for GEA-Efficiency to 2 per cent for GEA-Supply. GEA compares this with a projected growth in overall consumption of 200 per cent over the same period. However, that assumes that continuing economic growth is ecologically and economically sustainable, a belief that is questioned by several authors[5] as well as in this book. The high investment costs of low-carbon technologies would be offset, at least partially, by the savings in the costs of fossil fuels (which are not estimated by GEA) and by the removal of subsidies to the production and use of fossil fuels, which GEA quotes as about US$0.5 trillion and the International Energy Agency (IEA) estimates as US$1.2 trillion annually (including external costs) for the whole world (see Ch 9, 'Removing subsidies to fossil fuels and nuclear energy').

Another important conclusion of GEA was that, although technology is important, equally important are issues related to implementation: institutions, market behaviour, skills and capacities.

## Greenpeace's Energy [R]evolution

The Energy [R]evolution was commissioned by Greenpeace International, in association with the European Renewable Energy Council and the Global Wind Energy Council, from teams of expert consultants. Like the GEA report it divides the world into Organisation for Economic Co-operation and Development (OECD) regions and analyses them separately and together. In the reference or BAU scenario, the forecast of future energy demand to 2050 is based on expected future growth in population and gross domestic product. Technological change, away from fossil fuels to EE and RE, is then applied to obtain the Energy [R]evolution scenario, which is based on EE and RE. (This study assumes that economic growth is the same in both scenarios.) As a result demand peaks in 2020 and thereafter decreases slightly to the 2009 level by 2050. On the supply side, cost projections are estimated for RE technologies (decreasing) and fossil fuels (increasing). Since transport is an important part of this study and the most difficult one to transform to zero emissions, a projection of future oil production is also given, with peak oil reached around 2020. Fuel consumption and passenger-km travelled for motor vehicles are also projected.

Putting it all together, the result is that by 2050 almost all global electricity and heat are supplied by RE, but half of transport is still run on fossil fuels and there are still some non-energy GHG emissions from industry. Total $CO_2$ emissions decrease by 89 per cent from 28 gigatonnes (Gt) to 3 Gt per year, while in the BAU scenario they increase by 62 per cent. For electricity, in particular, during the first few decades the investment costs of the sustainable energy scenario are, as expected, much greater than the savings in fuel costs. But in the decade 2031–40 these paths cross over and by 2050 the accumulated costs of sustainable energy investments are outweighed by the savings in fuel costs (which include a carbon price).

## Electricity scenarios

Electricity production accounts for one-third of total global fossil fuel use and around 41 per cent of total energy-related $CO_2$ emissions.[6] Because it will be very difficult to supply large quantities of high-temperature industrial heat and urban transport in an ecologically sustainable manner by burning fuels, a likely major part of the ecologically sustainable solution for these forms of energy use is electrification. So, an increase in the proportion of energy delivered in the form of electricity is an important aspect of the pathway to a sustainable energy future, the Great Transition. Fortunately most of the technologies needed for a zero-carbon electricity future are already either commercially mature or at the early diffusion stage (see Ch 2). Therefore, a high priority, along with improving EE and implementing other ways to reduce demand at critical times, is to transform the electricity industry so that generation comes predominantly from zero net emission technologies as soon as possible.

Renewable electricity technologies work, are safe and are reliable. The reliability of nature's RE inputs to these technologies is discussed below. While renewable electricity technologies will still benefit from ongoing research, development and demonstration, a portfolio of these technologies is ready for rapid, widespread deployment. This will provide more operating experience, bring down costs, build the confidence of potential investors and simultaneously reduce emissions.

Global resources available for the generation of renewable electricity from the Sun and wind are huge, even after applying sustainability conditions (Ch 2). Although most RE power stations are currently more expensive than fossil-fuelled ones in terms of capital costs per unit of electricity generated, the prices of the former are decreasing with experience and expanding markets. Meanwhile their operating costs are, for the vast majority of RE technologies, less than those of fossil-fuelled and nuclear power stations. Keep in mind that the prices of the fuels used in conventional polluting power stations will probably increase in the long term.

Because the diverse RE sources are distributed inequitably over the globe, just as fossil and nuclear fuels are, different regions and countries will need different RE mixes. To some extent this distributional inequity can be reduced by trading renewable electricity, especially between neighbouring countries. For example:

- Under a low-demand scenario, Britain could generate most of its future electricity from wind and marine sources.[7] With higher demand, under the theoretical constraint of commercially available technologies, it does not have sufficient land or shallow coastal waters to become self-sufficient in RE.[8] However, it could possibly import solar and wind power from North Africa by transmission line via continental Europe.[9]

- In the USA[10] and Australia,[11] all or most electricity supply could come from wind, concentrated solar thermal (CST) and solar photovoltaics (PV). All the required technologies are commercially available.

- Japan, with its high population density and limited land area, has little space for on-shore wind farms and large solar power stations. However, it could have significant contributions from off-shore wind, rooftop solar PV and geothermal.[12] It may still have to supplement these sources by importing renewable electricity by transmission line from the Asian mainland or, in the long term, by importing solar hydrogen by modified liquefied natural gas (LNG) tanker from Australia.

- China and India have high potential for wind, PV, CST and, with large environmental and social impacts, hydro. However, until detailed resource assessment and scenario studies have been published, it is unclear whether these RE sources would be sufficient for phasing out coal.

With this background we can now address the principal issue to be resolved by electricity scenarios, the integration of large amounts of RE into the electricity grid. This is indeed the principal goal of most of the electricity scenario studies listed in Table 3.1.

## Electricity scenarios: computer simulations

The least expensive way to investigate the integration of large amounts of RE into an electricity network is to build a computer model of the system and simulate the operation hour by hour over years. The most credible models use actual hourly data on electricity demand and on relevant RE inputs at the same time-steps. The models postulate a mix of commercially available RE power stations appropriate to the country or region of interest and pass the RE data through the digitised characteristics of these power stations. The hypothetical generation mix may also contain some conventional power stations, whose operating characteristics are also fed into the model. An important characteristic of the model is the choice of the order in which power stations are dispatched into the grid, known as the merit order. This entails that the stations with the lowest operating costs, such as wind and solar PV, are dispatched first, and stations that use fuel, such as nuclear, coal, gas and biomass stations, are dispatched with lower priority, if and when they are needed to meet demand.

Each time-step, which is either an hour or a fraction of an hour, the model dispatches power stations in merit order; trying to balance supply and demand. At the end of a computer run through one or more years of data, the program summarises the results, for instance, the electrical energy generated per year by each technology; the average number of hours per year that supply failed to meet demand (known as the loss-of-load probability or LOLP); the corresponding energy shortfall; and the spilled or curtailed generation (that is, the energy supply that was excess to demand over the period concerned). Then, some of the initial assumptions are varied and more runs are performed. Of particular interest in such sensitivity analyses are the effects of varying the mix of RE sources, the merit order, the amount and types of storage in the system, and operating strategies for filling and using the storages. Since many things can be varied singly or several simultaneously, a single study may involve thousands of runs through the time period for which data are available.

With so many simulation runs the huge array of results may not point to any clear conclusions. To avoid this possibility, simulations are usually done with a goal, for example to find the mix of power station types that optimise the reliability of the generating system as measured by LOLP or energy shortfall, or to minimise spilled energy while meeting an existing reliability criterion. However, you may then ask, does it really matter if a mix of RE power stations spills a certain fraction of the energy generated when demand is low and the sun is shining brightly or the wind is blowing strongly? An even more important question is, how much storage and/or back-up is needed in a predominantly RE system? Ultimately these questions are about the economics of different mixes of power stations, different operating strategies and different types and amounts of storage.

The first step in exploring the economics of the RE transitions is to attach a set of costs to each power station in the model. There are capital costs that can be annualised by choosing an interest rate and economic lifetime; operation and maintenance costs; and, for some technologies, fuel costs. Since we are interested in a transition, we have to make assumptions about future cost trends. A good model will always set out its assumptions clearly. Since we don't want our computer model to output a large set of costs for each of thousands of runs through the time period of interest, the next step is to choose a method of optimising the economics, that is, finding the minimum costs subject to various assumptions.

In this account of computer simulations I have followed the

basic principle followed by many modellers: start with a simple model of a complex system, for which the results can be easily interpreted, and then gradually add complexity, making the model more and more realistic. (Climate models have evolved in this way.) Thus modellers build up their understanding of the system step by step and learn how each added complexity influences the results. As a member of an electricity system modelling team at the University of New South Wales (UNSW) and previously in the Commonwealth Scientific and Industrial Research Organisation (CSIRO), I'll now illustrate this method by outlining the UNSW approach to exploring the reliability and costs of hypothetical 100 per cent renewable electricity supply to the Australian National Electricity Market (NEM). Of course other modellers may have different approaches; indeed the combination of results from different approaches creates a robust understanding of a complex system.

## Simulations of 100 per cent renewable electricity

At UNSW we commenced modelling the integration of 100 per cent renewable electricity into the NEM in 2011. At that time those who were sceptical about the technical feasibility of running an electricity supply system predominantly on RE were claiming that, since RE is 'intermittent', it cannot provide electricity with the same reliability as fossil fuels or nuclear power. Therefore, the sceptics claimed, integrating large amounts of RE into electricity grids is only possible when it is backed up 100 per cent with conventional power stations. We were aware that a few studies had already refuted this claim in other countries with different electricity systems and decided to investigate this claim in the NEM, where electricity in 2010 was supplied by over 90 per cent fossil fuels, 77 per cent from coal, making it the third most greenhouse-intensive electricity supply in the world after Poland and South Africa.

Modelling is always constrained by available data. We obtained an excellent data set for hourly electricity demand, wind farm outputs and solar energy inputs, both Direct Normal Incidence (see Glossary) for CST and Global Horizontal Incidence (see Glossary) for PV, for 2010. So we decided to start simply: minimise our assumptions, use only commercially available RE technologies and explore the reliability of 100 per cent renewable electricity systems in 2010. Because Australia is blessed with a huge solar energy resource west of the Great Dividing Range and substantial wind energy potential in its southern regions, we initially chose a mix that would generate about 40 per cent of annual electrical energy from CST with 15 hours of thermal storage, 30 per cent from wind, 10 per cent from solar PV and the remainder as required from existing hydro and hypothetical biofuelled gas turbines. All are commercially available technologies, although the production of liquid and/or gaseous biofuels would have to be increased to fuel the gas turbines.

A key member of our team, Ben Elliston, a PhD student with considerable knowledge and experience in high-performance computing, designed a flexible computer program that simulated a whole year of hourly operation in a fraction of a second. This made it easy for us to perform sensitivity analyses on several of our initial assumptions. We found that 100 per cent RE could have supplied the NEM in 2010 with the same reliability as the existing polluting fossil-fuelled power stations. In the existing fossil-fuelled generation system the principal challenge is meeting demand on hot afternoons and evenings in summer, when the air conditioning demand is high, and a few loss-of-load events occur. In contrast we found that, in the 100 per cent RE system with half the electrical energy coming from the Sun, meeting the summer peak demands is easy. The principal challenge is meeting peak demands in winter evenings following calm cloudy days. Under these conditions the thermal storages are not full when evening comes, the gas turbines and hydro power stations are called upon, and some short (2-hour or less) loss-of-load events may occur. Nevertheless, over the whole year we could achieve the required reliability with peak-load back-up generating capacity much less than the maximum demand on the grid.[13]

In the sensitivity analysis we explored different ways of reducing the biofuel use and generating capacity of the gas turbines without loss of reliability. We could do this in several ways, for example by installing more CST; or increasing the area of solar collectors for the existing CST without increasing the capacity of the CST generators; changing the operating strategy for filling the thermal storage; increasing the geographic distribution of the wind farms,[14] and reducing demand during periods of low sunshine and wind. While increasing the CST contribution looked to be potentially a very expensive option, demand reduction was highly successful: a reduction of just 5 per cent in the winter peak demands removed all loss-of-load events. Such demand reductions have been practised for many years on a limited scale in Australia and several other countries. Aluminium smelters, which until recently took 13 per cent of Australia's annual electricity generation, are offloaded for up to two hours at a time when necessary, without interrupting the smelting process. With the current growth of 'smart' devices, it would be possible to offload a much wider range of electricity using appliances and equipment. For instance, air conditioners, refrigerators and freezers could be offloaded in a rolling program across the grid for a half-hour each without spoiling food.

The next step in the NEM simulations was to put prices on the construction and operation of power stations, to perform many simulations while varying the assumptions, and find the minimum annual cost. Initially we compared the economics of an economically optimal hypothetical 100 per cent RE system with a hypothetical new system of fossil-fuelled power stations with maximum efficiency of energy generation. As before, all technologies in the simulations were commercially available. We used 2010 data again for the simulations, while choosing the cost projections for 2030 of the Australian Energy Technology Assessment (AETA) performed by the Australian Government's Bureau of Resources and Energy Economics. (My personal view is that the AETA cost projections for RE are too high and for fossil and nuclear energy are too low.)

We found that, unlike the answer to the question of the meaning of life in *The Hitchhiker's Guide to the Galaxy*, there is no single simple answer. It all depends on the assumptions made. We did the economic simulations with two different discount rates (see Glossary) of 5 per cent and 10 per cent and two different cost sets, AETA's 'low' and 'high' costs, for RE.

In the optimal RE mixes, the cheapest technology, wind, makes the largest contribution to annual electricity generation, 46–59 per cent, followed by CST with thermal storage 13–23 per cent, solar PV without storage 15–20 per cent, gas turbines 6–7 per cent and existing hydro 5–6 per cent. Despite the huge total contribution from fluctuating sources, wind and solar PV (51–79 per cent in total), reliability is maintained. It turned out that this generation mix is similar to that found for 100 per cent global renewable electricity supply by Mark Jacobson from Stanford University and Mark Delucchi from the University of California at Davis: 50 per cent wind, 20 per cent PV, 20 per cent CST, 4 per cent hydro-electricity and 6 per cent from other sources.[15]

The comparison between the costs of the renewable and 'efficient' fossil scenarios is done by introducing a variable carbon price and observing the level at which the RE mix breaks even with the fossil mix. In the 5 per cent discount rate case, the RE system is less costly on an annualised basis when the carbon price exceeds the range $50–$65 per tonne of $CO_2$. With a 10 per cent discount rate, the RE system is less costly when the carbon price exceeds the range $70–$100 per tonne.[16] The higher discount rate disadvantages RE, whose costs are mostly capital costs. To compare these results with those of (say) European studies is difficult, because fossil fuel prices are much lower in Australia, while solar energy and wind speeds are higher.

In the absence of a carbon price, the annualised cost of the RE scenario is $7–$10 billion per year higher than that of the polluting 'efficient' fossil scenario. This is comparable with the estimated total annual subsidy to the production and use of fossil fuels in Australia of at least $10 billion.[17] Thus the additional cost of the renewable scenario could be paid by transferring the fossil subsidy to renewables.

Since an 'efficient' fossil scenario, based on commercially available technologies, is still very greenhouse intensive, it is not an appropriate comparison system in a climate friendly future. So our next step was to compare the 100 per cent renewable electricity system with hypothetical systems based on coal or gas with CCS. Because CCS systems are technologically unproven, we took wide ranges of future costs for CCS, the transport and storage of the captured $CO_2$, gas prices and carbon prices. We found that coal and gas with CCS scenarios will likely struggle to compete economically with 100 per cent renewable electricity, even if the technologies manage to be demonstrated and commercialised by 2030. Thus, in a climate-constrained world, coal-fired electricity may have little or no future in Australia, even if CCS eventually becomes commercially available.[18]

Many additional complexities need to be added to these simulations:

- The model of transmission systems needs to be made much more realistic and the choice of principal connections needs to be variable and included in the optimisation.
- Similarly, different types and amounts of storage should be incorporated in the optimisation.
- Further work is needed to extend the model to incorporate random variations in supply and demand, and hence the reserves required for such events. One way of doing this is to describe demand, conventional supply and RE supply at each data point by probability distributions instead of deterministic values, a method known as Monte Carlo simulation.
- Although hourly simulations over a single year are sufficient to establish the principal results, longer periods need to be studied to find any rare random events when, for example, there may be extended periods when there is little sunshine or wind. This is an alternative approach to Monte Carlo simulation.
- The effects of fluctuations in solar energy on timescales of

minutes, as the result of passing clouds, and the reduction of those fluctuations by means of a geographically dispersed set of solar power stations, needs further investigation.

Fortunately, much of the additional research is already in progress in a number studies around the world. For instance:

- Changes to the transmission network are included in the National Renewable Energy Laboratory (NREL) study on the USA.[19]
- The European simulations by Dominik Heide from Aarhus University in Denmark and colleagues[20] and Morten Grud Rasmussen from the Technical University of Munich in Germany[21] that study storage size and the reserves needed for balancing supply and demand, use hourly data spanning eight years.
- So far there are very few Monte Carlo simulations of the operation of a power system with high penetrations of RE.

Next I examine the scenario studies and computer simulations by research groups around the world to refute the three myths about RE listed in the introduction to this chapter.

## Myth-busting results and discussion

### Myth 1: RE is too diffuse or undeveloped

> Myth 1: Renewable energy is too diffuse or too undeveloped to run an industrial society.

While RE is certainly more diffuse, compared with fossil or nuclear energy, in terms of power production per square kilometre of land or water, that does not rule it out as the potential predominant source of energy for industrial society. It would make as much sense to insist that agriculture must be limited to dense production in feed-lots for animals and hydroponic sheds for vegetables. The relevant question is rather: 'Is there enough land and water surface area to generate the demanded energy from renewable sources in an ecologically sustainable way without competing with food production?' Chapter 2 shows that only a tiny fraction of the Earth's surface is needed to generate thousands of times more power than current demand from wind and solar alone. This is still true after applying sustainability constraints. Since not all regions of the world are equally blessed with RE sources, trade will be necessary. Trade can be conducted by transmission line or, for longer distances, by modified LNG tanker carrying solar hydrogen, as mentioned in 'GHG reduction scenarios: wide scope', above.

Another claim that is used by RE deniers to support this myth is that RE needs much more research and development before it can be considered as the principal energy source for industrial society. But this is simply wrong, as the discussion of RE technologies in Chapter 2 plainly shows. Solar-efficient design of buildings, solar hot water, on-shore wind power, hydro, electricity and heat from the combustion of biomass and several types of solar PV module are all commercially mature technologies in large-scale mass production (see Ch 2). Off-shore wind and CST with parabolic trough collectors and thermal storage in molten salt are commercially available in limited production – they are in the early diffusion stage of maturity. CST based on power towers and linear Fresnel collectors and several wave power systems are in the demonstration stage. Thus there is a substantial portfolio of clean energy technologies to choose from, depending upon geographic location. On the other hand, of the other technologies that are potential low-carbon contributors, coal power with CCS is just entering the demonstration stage and Generation IV nuclear power stations (fast breeder reactors) have been stuck at the demonstration stage for decades (see Ch 6).

The International Energy Agency (IEA), founded in response to the 1973–74 oil crisis, 'works to ensure reliable, affordable and clean energy for its 28 member countries'.[22] Over the decades its principal interests have been in fossil-fuelled and nuclear energy technologies. Because the IEA was slow to recognise the potential for RE, in 2009 supporters of RE set up an alternative organisation, the International Renewable Energy Agency (IRENA). However, even the IEA has published a low-carbon energy scenario, the Blue Map, in which RE provides nearly half of the global electricity by 2050.[23] While a scenario is not a forecast, the fact that the IEA allocated this contribution to RE may be interpreted as an expression of confidence in the technologies. Even greater confidence has been expressed by the Danish government, with an official target of 100 per cent renewable electricity and heat by 2035 and 100 per cent of all energy – electricity, heat and transport – to come from renewable sources by 2050.[24] Scotland, already with 35 per cent renewable electricity, has a target of 100 per cent renewable electricity for 2020. By then it plans to be exporting as much electricity as it consumes.[25]

Clearly, the myth that RE is too diffuse or too undeveloped to run an industrial society, is busted.

### Myth 2: RE is too unreliable

> Myth 2: Since renewable energy is intermittent, it is too unreliable to be the major source of electricity supply. Large penetrations of renewable energy into the grid must await either the development of base-load renewable power stations or a vast amount of storage.

Computer simulation studies with 80–100 per cent renewable electricity refute this traditional assumption. The simple conception of balancing base-load demand with base-load supply over a 24-hour period, discussed in Chapter 1 and illustrated in Figure 1.7, is hopelessly outdated in the context of an RE future. It was appropriate when inflexible base-load power stations, coal or nuclear, were the only choice for 24-hour power. However, when large penetrations of RE are integrated into the grid, base-load RE power stations are not needed. The computer simulations show that base-load demand can be supplied reliably by a mix of RE sources, none of which is, strictly speaking, a base-load power station, as shown in Figure 3.1.

It's more relevant to speak of a mix of 'flexible' and 'variable' RE power stations. The output of a 'flexible' station can be varied rapidly as required. Furthermore a 'flexible' technology is dispatchable, meaning it can be switched on rapidly and fed into the grid upon demand with a high degree of certainty.

The following renewable electricity systems are flexible to varying degrees:
- a bioenergy power station with an assured source of biomass fuel
- a hydro-electric station with a large dam
- a geothermal power station and
- to a lesser degree, a CST power station with a specified amount of thermal storage.

As with any definition, there are grey areas. The flexibility of an RE source with storage increases with the amount of storage.

FIGURE 3.1    **Meeting daily demand with flexible and variable RE sources**

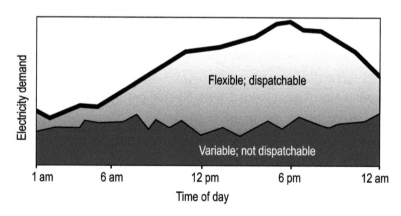

NOTE Compare the traditional, conventional conception in Figure 1.7.
SOURCE The author's adaptation of research results by David Mills and by Elliston, MacGill and Diesendorf.

On the other hand, wind, solar PV without storage and run-of-river hydro are much less flexible and are classified as 'variable' sources. Although the electrical outputs of variable power stations can be decreased when required, they cannot be increased beyond the level offered by weather or flow conditions at the time. In other words, they cannot be dispatched reliably, but that does not matter if they are part of a mix with flexible renewable power stations. Some people label variable RE technologies as 'intermittent'. I have avoided this term, because it can be misleading, implying incorrectly that the large-scale renewable electricity production generally cuts off abruptly when the wind drops or a cloud blocks the Sun. While the output of a single wind turbine or solar PV module may indeed be intermittent in this sense, a wind farm spanning tens of kilometres and a solar power station spanning a kilometre or so will take much longer to cut off. A system of geographically dispersed wind farms and solar power stations may take hours. Furthermore, short-term changes in wind and sunshine can be predicted quite well from weather observations.

The challenge for a predominantly RE system is to supply the peaks in demand, not base load, and this can be achieved with a mix of flexible and variable power stations. The concept of flexibility can be extended to the transmission network and to demand management as well, as shown by the results of detailed hour-by-hour computer simulations of supply and demand in the USA by the National Renewable Energy Laboratory (NREL), which finds that:

> … renewable electricity generation from technologies that
> are commercially available today, in combination with a more
> flexible electric system, is more than adequate to supply 80%
> of total electricity generation in 2050 while meeting electricity
> demand on an hourly basis in every region of the United States
> … [furthermore] increased electric system flexibility, needed
> to enable electricity supply-demand balance with high levels
> of renewable generation, can come from a portfolio of

supply- and demand-side options, including flexible conventional generation, grid storage, new transmission, more responsive loads, and changes in power system operations.[26]

In the 80 per cent renewable electricity scenario by NREL, about half the annual electricity generation comes from fluctuating wind and solar PV. The other renewable sources are CST, hydro, biomass and geothermal. Coal, gas and nuclear comprise the remaining 20 per cent, reduced to 10 per cent in NREL's 90 per cent renewable electricity scenario. Vast amounts of storage are found to be unnecessary for 80–90 per cent renewable electricity in the USA.

The results of scenario modelling are supported by observation. Two large-scale electricity systems already have more than one-quarter of their annual electricity generation coming from variable RE sources, predominantly the wind. In 2012 Denmark generated 30 per cent of its annual electricity supply from the wind.[27] Although Denmark is connected by transmission lines to Germany and Norway, it has so far made only modest use of these connections to smooth its fluctuating wind power. Contrary to the claims of RE critics, almost all Danish wind power is used in Denmark.[28] However, as Denmark expands its wind power capacity to meet its goal of 50 per cent wind energy generation by 2020, it will have to make increasing use of Norway's hydro storages and so a third transmission link between the two countries is under construction and plans for additional links to Germany and the Netherlands are being developed.

In the year ending June 2013, South Australia generated 27 per cent of its electricity from the wind, overtaking coal's contribution. In addition, 3.7 per cent of the state's electricity was supplied by residential PV.[29] As a result the Northern brown coal-fired power station (544 MW) is closed for six months of the year and another, smaller brown coal station, Playford, is likely to be shut down permanently. Although South Australia has only weak transmission connections to the neighbouring state of Victoria, it is planning additional wind farms, together with a modest increase in the capacity of the transmission link between the two states.

There are also several mini-grids in remote areas around the world where wind contributes more than 40 per cent of annual electricity generation in hybrid wind–diesel systems, for instance, at the Mawson Antarctic base and the isolated townships of Bremer Bay, Hopetoun and Coral Bay in Western Australia. Where the economics are favourable, as in locations where diesel fuel is very expensive, there is no technical barrier to very large contributions from wind.

So, while the variability of wind and solar PV offers challenges, it is not an insuperable barrier to 100 per cent renewable electricity.

### Myth 3: RE is too expensive

> Myth 3: Renewable energy is too expensive to be used on a large scale.

What is 'too expensive'? Global climate change, caused predominantly by the combustion of fossil fuels, is responsible for rising sea-levels, biodiversity loss and the increased frequency of extreme events, such as heatwaves, droughts, wild fires, floods and probably severe storms. The social and economic impacts on human health, agriculture, water resources, rural towns and coastal urban infrastructure of major climate changes, corresponding to global average temperature increases of 4°C or more, are likely to be huge.[30]

On the other hand, the costs of new RE sources are declining and some are projected to be competitive with fossil fuels within a few years or decades, given appropriate government policies during the transitional period (see Ch 2). Residential rooftop solar PV is close to breakeven with retail electricity prices in several parts of the world, while wind energy in high-wind regions is already cheaper than nuclear energy and may be cheaper than new gas and coal in southwestern USA and Australia.[31] With carbon prices that compensate for the environmental external costs of fossil fuels,[32] it can be argued that almost all commercially available RE technologies would be already economically competitive. While further cost reductions should be pursued through research, development and demonstration, this is not a good reason for delaying the rapid deployment of commercially available RE technologies, which will inevitably become cheaper as their sales increase.

## How rapidly can we transition?

Most of the scenarios mentioned in this chapter focus mainly on technological change. They suggest that we already have most of the technologies needed to dramatically reduce GHG emissions through demand reduction and RE. Studies also show that the transition, although initially expensive in terms of investment, is affordable. The principal barriers are the absence of effective policies to drive the transition.

Even the scenarios that recommend policies are concerned mainly with specific policies to encourage technological change, such as targets for reducing GHG emissions; targets for EE and RE; carbon pricing; feed-in tariffs for RE; regulations and standards for EE in buildings and appliances; and government funding for research, development, demonstration and new infrastructure such as railways and transmission lines. These and other policies and strategies are discussed in Chapters 8 and 9. However, avoiding serious and irreversible climate change requires rapid responses from the nations of the world, both collective and individual. Determining how rapidly the transition could be made is not primarily a technical problem. It involves broader aspects such as obtaining finance, training the labour-force, building manufacturing capacity and creating new institutions of governance, on global, national and sub-national scales. It requires a different kind of scenario study from most of the technical-economic studies described in this chapter. From those listed in Table 3.1, the German Advisory Council on Global Change (WBGU) study comes closest to addressing this need, at least on a global scale. This is an area where more research is needed.

Climate activists assert that rapid mitigation is feasible, invoking the scale and scope of wartime mobilisation strategies. To examine this notion more closely, UNSW PhD candidate Laurence L Delina draws upon historical accounts of social, technological and economic restructurings in several countries during World War II in order to investigate potential applications of wartime experience to radical, rigorous and rapid climate mitigation strategies. He finds that, while wartime experience suggests some potential strategies for rapid climate mitigation in the areas of finance and labour, it also has severe limitations, resulting from its lack of democratic processes. It seems unlikely that the industries and populations of nations would agree to being ruled by a 'war cabinet' to combat climate change. Restructuring the existing socio-economic system is more complex than fighting a war.[33]

## Conclusion

Compared with uncontrolled experiments on the real world, scenario studies are an inexpensive means of exploring the consequences of different assumptions about different future pathways. They provide visions for the Great Transition. There is already a large portfolio of peer-reviewed sustainable energy scenarios of various types for the whole world, regions and individual countries. They show that a future energy system, based predominantly on the efficient use of commercially available RE technologies, is technologically feasible. For electricity generation in particular, hourly computer simulations show that in many countries and other regions, supply from 80–100 per cent RE can be just as reliable as existing fossil and nuclear supplies.

At this stage only a small number of scenarios attempt detailed economic analyses of the transition. Although the details are still subject to uncertainties, the principal qualitative results are clear. On one hand the investment costs of a sustainable energy system are much higher than those of a fossil-fuelled system. On the other hand the operating costs of the sustainable energy system are much lower. Some economic analyses find that the savings in fuel costs (including carbon prices) offset all or a large part of the higher investment costs after a few decades. The range of results is not surprising, because the prices of some commercially available RE technologies are decreasing rapidly, while others that are still in limited production have large potential for price decreases as their markets and production scales expand in the near future. There is also uncertainty in the future trends in fossil fuel prices. In the long run, international prices for oil and coal are likely to increase, while prices for gas may decrease, as shale gas and coal seam methane become more widely traded.

Another element that must be taken into account in the economic analysis is that on a global scale the direct economic subsidies to the production and use of fossil fuels are still much greater than subsidies to RE (see Ch 9, 'Removing subsidies to fossil fuels and nuclear energy'). In addition, the failure of most governments to include the costs of environmental and health damage caused by fossil fuels in their prices is an indirect subsidy. A carbon price is one way of compensating for these subsidies to fossil fuels.

Science tells us that deep cuts in GHG emissions are needed urgently. Even a trajectory to achieve zero global emissions by 2050 may not be enough to avoid dangerous, irreversible climate changes. Yet in the scenario studies the time needed to make the transition is usually an assumption, sometimes backed up with arguments about the availability of RE resources in the region of interest, local manufacturing capacity, sources of finance and government policies. This timescale is uncertain, because it depends critically upon political will in international, national and sub-national spheres leading international agreements, national and sub-national government policies, political power structures and institutional structures. The role of citizens in influencing decision-makers to implement policies for rapid and effective climate mitigation is discussed in Chapter 11.

# PART B:
# WHICH TECHNOLOGIES ARE SUSTAINABLE?

# 4: SAVING ENERGY

Customers don't want kilowatt-hours; they want services such as hot showers, cold beer, lit rooms, and spinning shafts, which can come more cheaply from using less electricity more efficiently.

Amory B Lovins[1]

McKinsey & Company – a multinational consultancy working on major issues for senior management in business, government and other institutions – is seen as a pillar of the establishment. So many business managers and politicians were surprised when McKinsey published a series of reports showing huge potential – globally and in individual nations – for cutting greenhouse gas (GHG) emissions while saving money.

The now famous McKinsey cost curves show the amounts of GHG reductions and corresponding economic savings or costs of a wide range of energy efficiency (EE) and energy supply technologies and measures which mostly use renewable energy (RE).[2] See the simplified cost curve in Figure 4.1. In it you'll notice the total cost saving (the total area of all the bars below the horizontal axis) from EE is almost equal to the total cost (the total area of all the bars above the horizontal axis) of low-carbon energy supply. In other words, if we were smart, we could pay for most of the additional costs of RE through the economic savings from EE.

FIGURE 4.1    **Simplified cost curve for a low-carbon economy**

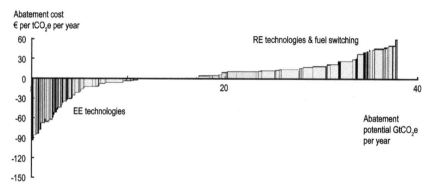

SOURCE The author.
NOTE The height of a vertical bar shows the net costs (positive or negative) of that individual technology; the width of a bar shows the reduction in GHG emissions it could achieve; and the area of a bar shows the total monetary cost (positive or negative) of implementing the technology.

Although the results and their illustration in cost curves had been well known to EE experts for decades, McKinsey's publication conferred respectability upon the field and damaged the credibility of economists who had been holding back progress on EE. Such economists had argued for decades that there could be no unclaimed economic benefits from EE because the alleged competitive market would have already claimed them – see further discussion in 'Barriers to energy efficiency', below.

The International Energy Agency (IEA) has been a pillar of the fossil fuel and nuclear industries. However, in recent years it too has become excited by the potential economic benefits and GHG reductions of EE. Its Efficient World Scenario

> … offers a blueprint to realise the economically viable potential of energy efficiency … Additional investment of $11.8 trillion in more efficient end-use technologies is needed, but is more than offset by a $17.5 trillion reduction in fuel expenditures and $5.9 trillion lower supply-side investment.[3]

The IEA scenario uses commercially available technologies. The key assumption is that policies are implemented to allow the market to realise the potential of all known EE measures that are economically viable. Despite projected growth in energy demand in non-OECD (Organisation for Economic Co-operation and Development) countries, global demand for both oil and coal peaks before 2020, but global demand for gas continues to grow through to 2035, the end-point of the scenario. As a result, global $CO_2$ emissions also peak before 2020 and return almost to the 2010 level of 30 gigatonnes (Gt) by 2035.[4] Although the EE technology measures would be insufficient to diffuse the climate crisis on their own, they could form the foundation of sustainable energy scenarios in which RE completes the emission reduction task. Indeed, unless growth in global energy demand is levelled off soon, it will become increasingly difficult and expensive to find sufficient good sites for renewable power stations to substitute for fossil fuels.

Like the IEA scenario study, the Global Energy Assessment (GEA) study introduced in Chapter 3, 'GHG reduction scenarios: wide scope' found EE to be the fastest and most cost-effective near-term option. It also found that EE has 'multiple benefits, such as reducing adverse environmental and health impacts, alleviating poverty, enhancing energy security and flexibility in selecting energy supply options, and creating employment and economic opportunities'. GEA offers a radical EE scenario that halves global energy use for heating and cooling of buildings by 2050. For industry GEA has a scenario that almost doubles global industrial output while increasing final energy demand by only 17 per cent.[5]

In this chapter we look at EE potential in buildings and industry, the two principal sectors of stationary energy use. (Efficiency in transport is included in Chapter 7.) Then we address barriers to EE, and strategies and policies for overcoming them.

But first we need to clarify the vague terminology that bedevils the field of energy saving. For instance, 'demand management' is often used as a generic term to refer to quite different concepts such as EE, energy conservation, overall reduction in demand for energy and shifting peaks in energy demand into troughs, a measure that may result in no actual reduction in energy use.

# Definitions

An *energy service* is a task or service that involves energy as an input; for example, home heating, office illumination, refrigeration, hot showers and transportation. The focus is on the service, rather than the quantity and type of energy supplied. Implementation of the energy service may need EE as well as energy supply. For instance, the energy service of a hot shower could be provided either with an electric hot water system and an ordinary showerhead or with a gas-boosted solar hot water system with a water-efficient showerhead. In regions where electricity is generated mainly from coal (eg, Poland, South Africa, Australia and China) the latter option has much lower $CO_2$ emissions.

*Efficient energy use* (sometimes simply called 'energy efficiency')[6] is using less energy to provide the same level of energy service – for example, by insulating one's home to use less heating and cooling energy to achieve the same ambient temperature; or installing fluorescent lights and/or skylights instead of incandescent lights to attain the same level of illumination; or replacing a fuel-wasting car with a fuel-efficient car of the same size, range and safety. Thus efficient energy use is achieved primarily by means of a more efficient technology or process rather than by changes in individual behaviour.

*Energy conservation* is using less energy to achieve a lesser energy service – for example, heating one's home less in winter without making improvements in EE; or driving less; or working in a less brightly lit room. A lesser energy service does not necessarily mean an inferior lifestyle: for instance, turning off some of the lights in an office may reduce illumination from a painful level of brightness to one that is comfortable, while still meeting work requirements; driving less may mean walking more and so becoming fitter and healthier. Thus energy conservation depends strongly upon individual perceptions, attitudes and behaviour.

As with other definitions, the boundary between efficient energy use and energy conservation can sometimes be fuzzy. For instance, a family that reduces its home heating in winter may wear warmer clothes. Is this using more efficient technology? To me, this is energy conservation, because it involves ongoing behavioural changes, rather than simply installing insulation in the roof and walls or a more efficient heating system.

Energy saving from efficient energy use and energy conservation has high value in both environmental and economic terms, especially when directed at reducing the consumption of electricity from fossil fuels. If the overall thermal efficiency of a coal-fired power station and transmission system is 29 per cent (see Figure 1.2), then one unit of electricity saved at the point of use actually saves about 3.5 units of primary energy from the combustion of coal. It is the primary energy consumption that determines the GHG emissions. Furthermore, saving one unit of electricity at the point of use in a home or office saves the retail price of electricity delivered to that home or office. Depending upon the electricity tariff, this could save 20–30 cents per kilowatt-hour (c/kWh), which is a much bigger economic saving than substituting for the price of electricity leaving the power station at 5–10 c/kWh.

Energy saving can be also conceived as *energy productivity*, getting more useful energy out of a joule or kilowatt-hour of energy. 'Productivity' has overtones of a business perspective with enhanced profits. Economists and governments place much emphasis on labour productivity, but strangely pay little attention to energy productivity, which should be at least as important.

*Peak demand reduction* is reducing the height of peaks in electricity demand. Even if it doesn't save energy (eg, if it only transfers a peak around 6 pm to a trough between midnight and 6 am), it can play a valuable role in facilitating the integration of fluctuating RE sources into the grid, as discussed in Chapter 3, 'Simulations of 100 per cent renewable electricity', and reducing the costs of the transmission and distribution system.

Next we define different degrees of EE improvement shown in Figure 4.2.

FIGURE 4.2    **Different degrees of energy efficiency improvement**

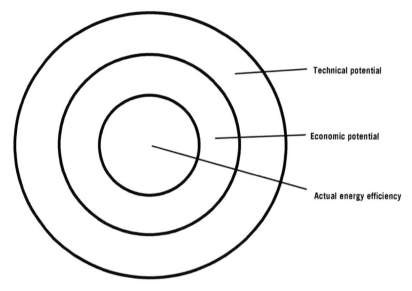

Technical potential

Economic potential

Actual energy efficiency

SOURCE The author.

The *technical potential* for EE is the maximum degree of EE that can be achieved for a particular level of energy service, irrespective of cost. In many cases, the technical potential for improving EE of energy-using appliances, equipment and industrial processes is huge. However, there is at present a large gap between the level of EE that is generally achieved in practice and the technical potential that is theoretically possible.

The *economic potential* is the degree of EE that is cost-effective at current energy prices and current interest or *discount rates*.[7] The economic potential is always less than the technical potential. Economic potential increases as the price of energy increases and/or the price of EE technologies decreases. Since the cost of EE is almost all capital cost, with a very low operation and maintenance cost, its economic potential increases as the interest rate or discount rate decreases. That is, the less we reduce or discount the value of future savings, the more these savings can outweigh the up-front cost of EE. Unfortunately, energy users tend to assign high psychological discount rates for investing in EE; for example, it is common to demand an economic payback period of five years or less, equivalent to expecting an annual return on the investment of about 20 per cent, which is much higher than could be gained from depositing the money in a bank. To a small extent, this reflects the high *transaction costs* (see Glossary) and inadequate information for these investments – in other words, market failures. However, to a large extent this represents behaviour that is economically irrational, since many EE measures are very cheap and a payback period of even 10 years is generally a good investment. For comparison, the developers of large power stations tend to apply a relatively low discount rate, equivalent to an economic payback period of 20 years of more, when deciding to build a new plant. This is economically irrational in the opposite direction, because the financial risk of borrowing money to build a conventional fossil-fuelled power station in a $CO_2$-con-strained world is not as yet adequately reflected in the interest rate charged.

It is widely agreed among energy consultants that the actual energy efficiency of the economies of many countries, both highly developed and less developed, falls far below the economic poten-tial, let alone the technical potential. The potential is illustrated in Figure 4.3, which compares trends in electricity consump-tion in California and the USA as a whole. Among the US states California has been exceptional in terms of its EE programs: programs for efficiency in electricity generation commenced in 1976 and for EE in buildings and appliance standards around 1977. The graph indicates the EE was responsible for part of the difference between California and the USA – other factors, such as climate and the growth of RE, would also have contributed. By the way, some energy policy people call the actual energy efficiency the *market potential*, but that seems inappropriate, because market failures are endemic in the field of EE, as discussed in the next section, and so 'market potential' is meaningless. It all depends on policies.

FIGURE 4.3    Trends in electricity use: California versus whole of USA

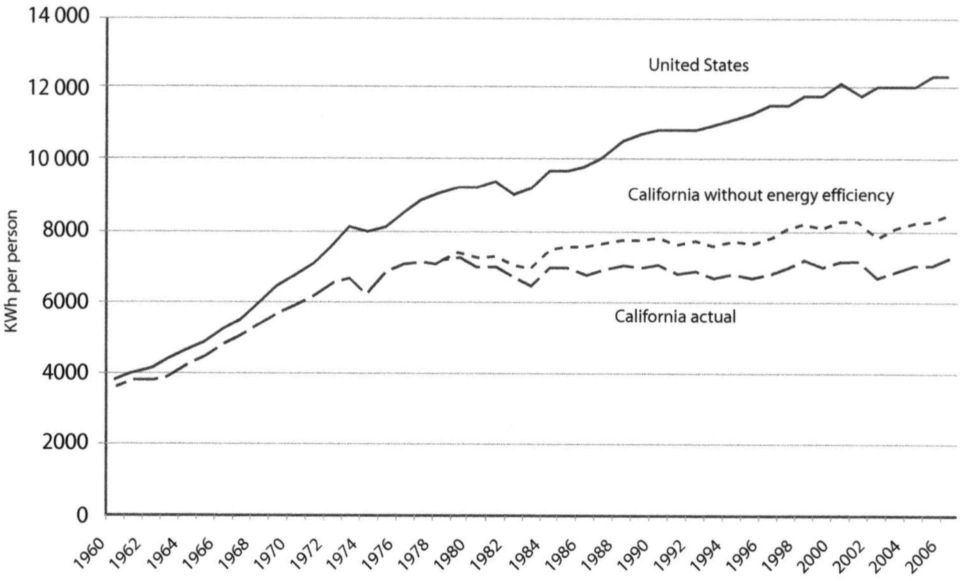

SOURCE Rosenfeld and Poskanzer (2009).[8]

## Energy efficiency: buildings and industry

### Buildings

Buildings and energy-using appliances and equipment inside the buildings are together responsible for about 30 per cent of global energy end-use and about one-third of energy-related $CO_2$ emissions.[9] Many solar-efficient (that is, EE plus passive solar design) buildings with very low energy consumption are scattered around the world, but there are not many whole suburbs or towns designed to be solar efficient. This demonstrates a gap in facilitating policies by national and state/provincial governments. So far municipal and local governments have the best records in several countries.

While the largest levels of EE can be achieved at least cost by excellent design and appropriate orientation of new buildings, there is also substantial potential for retrofitting many existing buildings. For both new and existing buildings the additional capital cost of many EE investments is paid back from the savings in energy bills, although the payback period for substantial energy savings is generally longer for retrofits.[10]

Inside energy-inefficient residential buildings in developed countries, the principal energy uses are generally for space heating, hot water and electrical appliances, especially refrigerators. Key technologies for improving EE and reducing $CO_2$ emissions are:

- large windows on appropriate walls to increase solar input in winter and eaves or deciduous trees, vines and bushes to exclude the summer sun
- insulation of the building envelope and hot water tank
- draught exclusion in winter and increased ventilation in summer
- electric fans or evaporative coolers
- solar or electric heat pump water heating and
- energy-efficient appliances and equipment.

A detailed discussion of these technologies and the practicalities of their use are beyond the scope of this book. However, the Appendix offers suggestions on what you can do to make your home and other activities more energy and $CO_2$ efficient. Many books and magazines address practical aspects.[11] Behavioural change is as important as EE in the residential sector.

It should also be noted that prematurely demolishing and replacing functional buildings and discarding and replacing functional appliances and equipment, all in the name of EE, has its costs in terms of $CO_2$ emissions, since existing buildings etc have embodied energy (the energy required to obtain raw materials, construct or manufacture the building, and to manage the wastes they produce).

The principal energy uses in energy-inefficient commercial buildings are generally heating, ventilation and air conditioning (HVAC), lighting, electronic equipment and, in the case of retail, refrigeration. These are easily addressed by energy service companies (see 'Overcoming barriers: strategies and policies', below). Poor building design is not easily changed in existing buildings.

Taking all buildings together, GEA's scenarios suggest that global building energy use for heating and cooling could be halved by 'the proliferation of today's best practices in building design, construction, and operation, as well as accelerated state-of-the-art retrofits'. It calculates that this achievement would require about $14 trillion in additional undiscounted cumulative investments to 2050, which would return benefits of about $58 trillion in undiscounted energy cost savings over the period.[12]

New buildings and precincts that use zero net energy in their operation, or are even net energy producers, are possible and some demonstration buildings and communities exist. However, this is only achievable for particular building types and settlement patterns, mainly low-rise buildings and less densely populated residential areas. To reduce transport energy use, medium- and high-density settlements are preferred (see Ch 7). It should be noted that even a 'zero energy' building requires energy to provide its components and material contents, and this is provided by industry.

### Industry

All industrial processes need energy to convert raw materials into products and so it is not surprising that industry is the other big contributor to end-use energy. Like buildings, industry is responsible for roughly 30 per cent of global end-use energy on average, although variations between countries can be very large.[13] The most energy-intensive and greenhouse-intensive industries are aluminium, iron and steel, cement, pulp and paper, and chemicals. Some of these industries have sources of GHG emissions additional to those from their huge energy uses:

*   Steel uses metallurgical (coking) coal both as an energy source and as a reducing agent for iron ore and both processes emit

$CO_2$. Biomass grown and collected sustainably could substitute partially for coke (see Ch 5, 'Water and other materials').

- Concrete based on Portland cement emits $CO_2$ when it sets. A new type of cement called Eco-cement absorbs $CO_2$ when it sets, but it has not yet been accepted by the industry (see Ch 5, 'Water and other materials').
- In aluminium smelting, perfluorocarbons (PFCs) with global warming potentials (see Glossary) per molecule of several thousand times that of $CO_2$ are emitted. The industry is working to reduce these emissions.

For improving EE in industry, substantial gains with rapid economic paybacks could be made in several widely used technologies – such as electric motors, boilers and kilns – as well as in whole systems of production. However, industry will always remain a significant energy user, so replacing its use of fossil energy with RE is essential to complement the EE improvements and to cut emissions.

## Barriers to energy efficiency

A few neoclassical economists (see Glossary) will tell you that there is no potential for additional cost-effective EE, because the market would have already implemented any opportunities. Their favourite illustration is 'You don't find $50 bills lying in the street, because if someone drops one, someone else will pick it up almost immediately.' These economists place blind faith in simplistic economic theory, while ignoring the reality that extensive market failures undermine attempts to implement EE. An important major type of market failure is known as the *split incentive*. Here are three examples:

- In the absence of mandatory energy performance standards for buildings, landlords aim to minimise their costs and, as a result, fail to make their property energy efficient. However, tenants, who usually pay the energy bills, cannot reduce their energy costs substantially without making changes to the building envelope, which would benefit the landlords.

- Any technology or measure that reduces sales of electricity from the grid – such as EE, solar hot water systems and residential solar photovoltaic (PV) systems – leads to revenue losses for the electricity utilities, even though these measures may be highly cost-effective for their customers. So, in the absence of appropriate government policies, utilities are unlikely to go beyond token gestures.
- In many countries, consumers are protected from excessive rises in electricity prices by a cap on the price per unit of electricity sold, which further encourages electricity wholesalers and retailers to sell as many units of electricity as possible, irrespective of environmental concerns.

Another type of market failure arises because small, specialised EE businesses generally have to compete with large energy supply utilities. These utilities are the principal customary suppliers of energy to consumers and so have several advantages in the market: regular communication with their customers; databases on their customers' patterns of energy use; the right to visit customers' premises to read meters; and often the goodwill of their customers. In the absence of government policies, the utilities themselves are unlikely to assist their customers to become energy efficient, against the utilities' business interest.

Yet another market failure results from the fact that almost all the cost of EE technologies is capital cost that has to be paid up-front. Even when a technology is highly cost-effective with a rapid repayment rate, many low-income energy users simply cannot afford to implement it. Government policy is needed to assist low-income earners to implement EE, avoid energy poverty and reduce emissions. Weatherisation programs, and incentives and regulations to require landlords to make their buildings energy efficient, are needed.

Many energy users lack information and expertise about EE.

Even where there are unbiased sources of information, the task of comparing sustainable energy technologies that have high capital costs but low operating costs with traditional energy supply technologies that have lower capital costs but higher operating costs, is difficult and time consuming for most people. To make matters worse, only a minority of electricians and plumbers are trained in the installation of EE technologies. Lack of quality control of EE products and installation further increases the difficulties for consumers.

The small scale of EE technologies creates another barrier. EE is usually excluded from many of the incentives available to large energy suppliers, such as tax deductions, loan guarantees, infrastructure development by government and infrastructure bonds. Also, compared with energy supply technologies, EE is not a fashionable area of research and so progress in improving products has been slow.

For a household or business, making a building and its contents energy efficient involves researching a wide range of products and services, for example, on insulation; window treatments; lighting; choice of heating, cooling and air conditioning systems; and choice of appliances and equipment. The task may involve many internet searches and phone calls, followed by negotiations with many different specialist tradespeople. Economists refer to these barriers to EE as a 'transaction costs', arguing that the additional inconvenience of purchasing these many small-scale goods and services represents real unavoidable costs that must be factored into the costs. Although this is true, it is not helpful, because it suggests incorrectly that these barriers are fixed and immutable. In reality, governments can remove them simply and inexpensively by appropriate policy changes.

## Overcoming barriers: strategies and policies

Strategic thinking about energy saving starts with the recognition that existing energy institutions and policies are biased towards energy supply, despite the huge short-term potential of reducing unnecessary demand. There is also a bias, not limited to the energy sector, towards making small improvements at the margin. In the face of the climate crisis, this is a totally inadequate situation. Therefore, before listing specific policy recommendations to promote EE, we consider the following broad strategies for addressing these bias barriers: systems thinking, radical whole system change, energy service companies and reducing rebound.

### Systems thinking

Substantial improvements in EE arise from re-examination of the whole system that delivers an energy service. Standard analyses focus on single elements, such as an electric motor. In a broader conception, energy consultant Alan Pears gives a simple example of a system where an electric motor drives a pump that circulates a liquid around an industrial site.[14] The system comprises six principal elements: electric motor (characterised by its size and efficiency); motor controls (switching, speed or torque control); motor drive system (belts, gearbox, and so on); pump; pipes; and demand for the liquid (or, in many cases, the heat or 'coolth' it carries). For simplicity, consider the simple hypothetical situation in which the efficiency of each element is improved by 10 per cent, and the efficiency of each element is independent of the efficiency of the others. (The real situation would be more complex.) Then the overall level of energy use, compared to an original level of one unit, is:

$$0.9 \times 0.9 \times 0.9 \times 0.9 \times 0.9 \times 0.9 = 0.53.$$

That is, energy savings of 47 per cent are achieved overall. Pears points out that in practice such a systems perspective is rarely applied, because responsibilities for different elements are allocated to different groups. This situation reveals the importance of creating appropriate organisational structures as well as improving hardware and software.

## Radical whole-system change

Let's take systems thinking even further, to the extent of redesign-ing the whole system. This can overcome the barrier to increased EE resulting from thinking in terms of small improvements at the margin. The books *Factor 4* and *Natural Capitalism* give exam-ples of very large improvements that can be achieved by rethink-ing, redesigning and re-engineering the whole system to produce particular energy services.[15] Their examples involve commercial buildings, electrical appliances, office equipment, windows, fans, motors and cars. They achieve 'Factor 4' improvements, which means that they reduce inputs of energy and materials to provide a given unit of service by a factor of four (75 per cent).[16] Nowadays there is even a Factor 10 Institute, which is exploring means of achieving reductions by a factor of ten (that is, 90 per cent) in the use of energy and materials while delivering the same service.[17] Needless to say, policies to make low-cost finance available are essential for tackling whole system change.

## Energy service companies

Energy service companies (ESCOs), sometimes called 'energy performance contractors', are businesses that sell energy services rather than just particular forms of energy.[18] With the assistance of an ESCO, energy users who wish to make their homes, offices and factories more energy efficient do not have to make separate arrangements with suppliers and installers of insulation, efficient lighting, solar hot water, efficient heating and cooling systems and industrial equipment. The ESCO is a one-stop shop that organises the whole package, with low transaction costs for customers, just like a kitchen renovator who does the whole kitchen. The ESCO identifies and evaluates energy-saving opportunities and then rec-ommends a package of improvements to be paid for through the energy savings achieved. The ESCO guarantees that savings meet or exceed annual payments to cover all project costs – usually over a contract term of 7–10 years. If the planned savings do not materialise, the ESCO pays the difference, not the customer. To ensure savings, the ESCO offers staff training and long-term maintenance services.

ESCOs in the USA, Europe and Australia are doing excellent business assisting large energy users to use energy more efficiently. However, the industry needs independent measurement and verification of energy savings. The projects are mostly in the commercial and industrial economic sectors where the projects are easiest and the profits are highest. ESCOs that assist individual households to become more energy efficient across the board have struggled to succeed due to high transaction costs and, until recently, low energy prices. However, transaction costs for households could be reduced by treating a large number of households as a single unit, obtaining baseline data from electricity and gas bills and then applying a package of low-cost measures, such as those listed in the Appendix. This method could be implemented by government housing departments for public housing and by ESCOs for private housing.

### Reducing 'rebound'

An apparent barrier to EE, resulting from a narrow conception of the problem, is the notion of the 'rebound effect', which is widely promulgated by some neoclassical economists. This effect arises from the situation that people who save energy usually save money as a result. Then the economists assume that the money saved is spent on products that increase energy consumption again: the 'rebound'. A particular case that has been observed is that people who insulate their homes subsequently come to expect a higher level of thermal comfort. Before insulating, they heated individual rooms while the rooms were occupied. After insulating, they may heat the whole house continuously.

The mental image created by 'rebound' is of a rubber ball that is dropped and rebounds back to the same height at which it is released. However, in reality, the size of the rebound depends on whether the money is invested in energy wasting (eg, a plasma television) or energy saving (eg, a new 5-star refrigerator to replace an old 2-star) products and services. On average, in the absence of EE policies by government, we would expect that people would spend the money in the same proportion as energy's proportion of gross domestic product (GDP), typically 8–10 per cent in industrialised countries. The IEA assumes 9 per cent in its Efficient World Scenario.[19] This is quite a small rebound.

The height of the rebound can be further reduced by government policies to promote EE through mandatory energy performance standards, energy audits and labelling, carbon pricing, education and information (see below). Furthermore, the rebound could be eliminated entirely by shaping the market for energy services so that it delivers to consumers packages of EE and RE in which the economic savings from EE pay for all or most of the additional costs of RE, as illustrated in Figure 4.1. Then huge reductions in $CO_2$ emissions would occur, but there would be little or no additional money to spend on rebounds. Indeed, it is even possible to envisage scenarios in which people could be nudged to invest money saved from an EE improvement in a further EE improvement.

A different kind of 'rebound', that is much more difficult to stop than rebound in the energy sector, is the result of macroeconomic growth. A radical solution is transitioning to a steady-state economy, discussed in Chapter 8, 'Changing the economic system'.

## Policies

Many EE technologies are mature and highly cost effective. Unfortunately, as discussed in the previous section, they are inhibited by several market failures and biases. So government regulations, incentives and disincentives must play important roles in fostering EE. Even so, a diverse portfolio of policies is needed to remove these barriers. The following policies for governments would set the framework for action by businesses and the general public:

- Targets and a strategy for achieving them are essential.
  For example, the European Union has a target to reduce
  primary energy consumption by 20 per cent by 2020
  compared with projections. It also requires its members to
  develop National Energy Efficiency Action Plans.[20]
- Mandatory energy ratings, energy labelling and energy
  performance standards are necessary for all buildings, new and
  old. Denmark commenced this process when the Organization
  of the Petroleum Exporting Countries (OPEC) quadrupled
  oil prices during the 1970s. Denmark ensured that EE
  improvements would be long-term by maintaining the high
  oil price by taxing it after the OPEC prices had fallen.
  In 2012, Denmark's goal of 100 per cent RE (electricity, heating
  and transport) by 2050 is an important motivator for further
  increasing its EE. Although its buildings are already very
  energy efficient compared with those in the USA or Australia,
  its latest short-term EE goal is a 7.6 per cent reduction in
  primary energy consumption by 2020.[21]
- An institution should be set up to provide homeowners on
  low incomes with low-interest loans for EE improvements
  and fuel switching of space and water heating from fossil
  fuels to solar heating and electric heat pumps. Alternatives
  are weatherisation or weather-proofing programs, in which
  governments undertake the task of making the homes of
  low-income earners more energy efficient.[22]
- Mandatory energy ratings, energy labelling and energy
  performance standards are essential for all energy-using
  appliances and equipment.
- In addition, governments could charge feebates (additional fees)
  on the sale of moderately energy-efficient appliances that just
  meet the mandatory standards and transfer the revenue raised
  to fund rebates on highly energy-efficient appliances.
- Energy information offices are needed in all cities and on-line to
  reduce the transaction costs for people who wish to make their

homes and businesses more energy efficient. Unfortunately many existing energy information offices were closed when governments corporatised or privatised public energy utilities.

- The tension between implementing EE and increasing sales of energy can be resolved by transforming utilities from being energy suppliers to becoming ESCOs. Governments can facilitate this by placing caps on the revenue that public utilities may earn from energy sales. Unfortunately this runs counter to the current dominant paradigm of fostering energy markets. Utilities should be permitted to charge for assisting customers to save energy.
- Governments could also give incentives to ESCOs to expand the scope of their businesses to the residential and small business sectors and the more challenging EE improvements in industry.
- All new industrial equipment and processes must meet mandatory EE standards.
- Government incentives are needed to encourage the early phase-out of energy-inefficient industrial equipment and processes.
- Mandatory EE standards, higher than those for the general public, should be set for buildings, appliances, equipment and motor vehicles purchased, leased or rented by government departments and agencies. Thus the market for high EE products can be expanded.
- Existing industry incentive schemes, such as tax write-off rates and loan guarantees, should be modified to favour industry investment in EE products.
- Governments should facilitate the application of infrastructure bonds to provide finance for EE projects that pool together large numbers of small, dispersed EE projects.
- Governments should direct more funding for technical and further education specifically towards energy audits and EE technologies.
- Research and development (R&D) funding from governments

should reserve a portion of investment for EE.

- Official measurements of EE improvements should be in terms of energy use per service. On a national scale 'energy intensity' (that is, energy use per unit of GDP) is not an appropriate measure of EE, because it decreases whenever GDP growth is greater than energy growth. Thus it can give the false impression that EE is improving when in some cases it is not.
- Government funding is required for 'smart grids' and 'smart devices' to foster demand reductions. This should be complemented by contracts between electricity suppliers and their customers that specify the maximum demand received by the customer. The demand component of the bill would be proportional to the contracted maximum demand, as discussed in Chapter 9, 'New business models for electricity industry'.

In addition to the leadership needed from governments, there are opportunities for community groups to take initiatives with group purchases of EE goods and services.

## Conclusion

The largest and fastest short-term potential for cutting GHG emissions comes from energy saving, the demand side of the Great Transition. This involves the implementation of policies to research, develop, demonstrate and disseminate EE technologies; to encourage energy conservation by behavioural changes; and to reduce demand with the assistance of 'smart' grids and 'smart' devices. The benefits of energy-saving programs also include greater energy security, less poverty, more local jobs, less air and water pollution, and better health.

The present system is biased away from energy saving towards energy supply. This is even true within the broad field of sustainable energy.[23] EE and RE are complementary and must be further developed and disseminated together. To level the playing field for energy saving, the focus of strategy, policy and implementation must shift from energy supply to energy services. This involves changes to institutions as well as to targets, regulations and standards, finance, pricing, grants for R&D, subsidies and government procurement. In particular, the growth of ESCOs should be facilitated and their scope broadened to have a bigger impact in the residential sector.

While marginal improvements are often useful, they can sometimes act as diversions or displacement activity with minor benefits when the major benefits come from radical whole system change. This challenges the conventional processes of government and business.

# 5: RENEWABLE ENERGY TECHNOLOGY IMPACTS

> Health complaints [about wind turbines] were as rare as proverbial rocking horse droppings until the scare-mongering groups began megaphoning their apocalyptic, scary messages to rural residents.
>
> Simon Chapman[1]

I frequently find myself on platforms before diverse audiences: academic, professional, business, environmental, trade union, political and faith groups, and interested members of the public at large. While the vast majority of questions and comments are perceptive and constructive, a minority reveal that people have been fed myths unfriendly to renewable energy (RE). These myths circulate through the community and are fostered by dedicated RE deniers and some sections of the media. While I have discussed some of these in Chapters 2 and 3, this chapter examines and refutes new myths about RE technology (see Table 5.1). Each myth has been busted by several independent research groups. Results of scientific and technological research should be shared widely, so people can make informed decisions, and governments can lead with policy based on reality, not fed by fear and misconceptions.

In this chapter I address the resource, environmental, health, social and economic impacts of RE that have not been treated in Chapters 2 and 3. In doing so, I mop up the remaining myths listed in Table 5.1.

TABLE 5.1    **Common myths about renewable energy**

| Myth | Location of refutation |
| --- | --- |
| 'RE is too diffuse. There isn't enough land to run an industrial society on renewable energy.' | Ch 2, 'Renewable energy resources' and Ch 3, 'Myth-busting results and discussion' |
| 'RE isn't mature enough to replace fossil fuels.' | Ch 2, 'Maturity of energy technologies' and Ch 3, 'Myth-busting results and discussion' |
| 'RE is too unreliable. A large-scale electricity supply system needs base-load power stations and RE can't provide it.' | Ch 3, 'Myth-busting results and discussion' |
| 'RE is too expensive to be used on a large scale.' | Ch 3, 'Myth-busting results and discussion' and Ch 5, 'Quantified costs and unquantified benefits' |
| 'More energy is needed to build solar PV modules than they can generate in their lifetimes.' | 'Energy payback', below |
| 'Wind turbines make people sick, even at a distance of 10 km.' | 'Other environmental, health and social impacts', below |
| 'All bioenergy systems have huge environmental impacts.' | 'Other environmental, health and social impacts', below |

## Energy payback

Under what circumstances would we collect fuels or build energy conversion systems for which the energy inputs are greater than the lifetime energy outputs of the fuel or system? Well, it makes a kind of sense for some purposes, such as industrial food production, in which energy inputs from the use of farm machinery and fertilisers are in total about ten times the somatic energy output from the food. Another example comes from the early use of solar photovoltaic (PV) modules to power human-made satellites. In those days, before PVs were mass produced, the energy input to manufacture them was greater than their electrical energy output. At that time it made sense for that specialised purpose. Nowadays solar modules are becoming an important energy source on Earth and they generate typically 10–25 times the energy required to manufacture them (see below).

It may also be reasonable to use a lot of low-grade energy in order to produce a little high-grade energy, provided the low-grade energy input is renewable and doesn't require an expensive and energy-intensive collector system. An example, outlined in Chapter 2 'Hydro-electricity and marine energy', is the generation of renewable electricity from the small difference in temperature between the ocean surface and deeper ocean produced by sunshine. However, in general it would be pointless to undertake coal mining, oil and gas recovery, nuclear power and any large-scale RE generation system in which the life-cycle, high-grade, non-renewable, energy inputs were greater than the outputs. By 'life-cycle energy inputs' I mean all the energy inputs from the mining, milling and processing of the raw materials, through the construction, operation and dismantling of the energy conversion system, to the management of the wastes. As the 'easy' reserves of fossil fuels and uranium are being used up, more 'difficult' sources are sought (eg, oil drilling in the deep ocean; mining low-grade uranium ore) and the energy inputs are increasing.

Even more important in terms of environmental impacts is the source of the energy inputs, that is, whether it is fossil or renewable. If we wish to reduce GHG emissions, there is little value in generating RE under circumstances where the $CO_2$ emissions from fossil energy inputs are comparable with – or indeed greater than – the emissions saved by running the RE generator instead of a fossil-fuelled generator. This is sometimes the situation with the production of biofuels, as discussed in 'Other environmental, health and social impacts', below.

A useful indicator of net energy generation of a technology is the energy return on investment (EROI), which is simply its lifetime energy output $E_{out}$ divided by the life-cycle energy inputs $E_{in}$.

$$EROI = E_{out} / E_{in}$$

(Equation 5.1)

Natural RE inputs – such as the solar energy used to grow biomass used to produce bioenergy – are not usually counted in $E_{in}$. The bigger the EROI, the better. If the EROI is less than one, the technology is a net energy sink rather than a source.

Another useful indicator is the energy payback period ($T_E$), the number of years that an energy conversion system must operate in order to generate its total life-cycle energy inputs. It is equal to the lifetime of the technology $t_{life}$ in years multiplied by the ratio $E_{in}$ / $E_{out}$. Thus

$$T_E = t_{life} \, (E_{in} / E_{out}) = t_{life} / (EROI)$$

(Equation 5.2)

The smaller the energy payback time, the better. For example, for a good quality solar PV system, the lifetime $t_{life}$ may be 25 years. Then if the lifetime energy output $E_{out}$ equals 25 times the life-cycle energy input $E_{in}$, the energy payback period is 1 year, which is being achieved now for some solar PV modules. On the other hand, if $E_{out} = E_{in}$, then $T_E = t_{life} = 25$ years, which was the situation back in the 1980s when solar PV produced no net energy over its lifetime.

Actually, the calculation of $E_{in}$ is not that simple, because decisions have to be made as to where to make a cut-off in what could be an infinite set of inputs. Do we include the energy required to make the machines that mined the raw materials that formed the energy conversion system of interest? A rigorous life-cycle analysis identifies a point where the additional inputs become negligible and cuts off the calculation there.[2]

The whole process of calculating the life-cycle $CO_2$ inputs becomes much easier as RE makes a bigger and bigger contribution to the national and global energy systems. Then RE conversion systems can be manufactured predominately by RE and they become RE breeders. Already some factories that make PV modules are powered either directly by rooftop PV modules or by purchases of 'green' electricity from a wind farm. For the long term there is an ambitious project proposal called the Sahara Solar Breeder in which factories would be built to manufacture silicon from the desert sand, solar PV cells from the silicon and solar power stations from the PV cells. Some of the electricity from the solar power stations would be used to expand the manufacturing of silicon and solar cells until enough electricity could be generated to supply all Africa and possibly the whole world.[3]

Table 5.2 gives approximate energy payback periods for some energy generation technologies. The range of values given for some technologies takes into account that energy outputs vary with location and fuel quality, if there is a fuel. For instance, a wind turbine at an excellent site may have an energy payback period of 3 months, but the same turbine at a moderate wind site may take 9 months to generate its energy inputs. A silicon crystal PV module at an average US site may require 1.7 years while the same module in the sunny south-west USA may achieve 1.2 years. The short energy payback periods for wind and solar PV makes RE breeders a feasible proposition for rapid climate mitigation without producing a huge spike in GHG emissions.

TABLE 5.2  **Energy payback periods for various technologies**

| Technology | Energy payback period (years) | Expected lifetime of technology (years) | Reference[4] |
|---|---|---|---|
| Solar PV modules | 0.5–1.8 | 25 | Fthenakis (2012) |
|  | 1.5 | 30 | Raugei et al. (2012) |
| Wind turbines (large) | 0.25–0.75 | 20–25 | Martinez et al. (2009) |
| Concentrated solar thermal (CST) , parabolic trough | 2 | 30 | Desideri et al. (2013) |
| Nuclear (high-grade uranium) | 6.5 | 30 | Lenzen (2008) |
| Nuclear (low-grade uranium) | 14 | 30 | Lenzen (2008) |
| Coal: conventional | 1–2 | 30 | Raugei et al. (2012) |
| Coal with CCS | ? | ? | ? |

NOTES
(a) Results for RE technologies vary sensitively with location.
(b) Some of the above results are contested in the scholarly literature on the basis of method.

## Water and other materials

With the growth of population, industry and large-scale agriculture, fresh water is becoming a scarcer resource. Although the total quantity of water is conserved, it becomes almost inaccessible for essential purposes such as food production when it is polluted, or evaporated, or extracted at a high rate from natural artesian storages by human activities. The vast majority of coal-fired and nuclear power stations use huge quantities of water, often fresh water, for cooling.[5] In some places such as Australia, households were rationed severely during the recent 10-year drought, while nearby power stations were using up (evaporating) drinking water at a rate comparable with that of a small city and were paying almost nothing for it. It is possible to cool power stations using air rather than water, but this adds significantly to the cost of electricity and so it is not common practice for coal power and is not done at all for nuclear power.

Most RE sources use very little water. Even the water that passes through a hydro-electricity power station is still available for irrigation and drinking downstream. A temporary exception to the general rule is concentrated solar thermal (CST), which uses water for cooling and for cleaning its mirrors. However, since many future CST power stations will be located in sunny but dusty desert regions, a solution is needed. Water cooling can be replaced by air cooling – for a price. For cleaning the mirrors one possibility is to commercialise the system used to clean the solar PV panels powering the Mars Exploration Rovers. The panels have a special coating on their surfaces. From time to time pulses of electric current from the Rovers' batteries are passed through the coating, dislodging 90 per cent of the Martian dust within 2 minutes. The energy used in these jolts is much less than the energy collected.[6]

Life-cycle analysis can also keep track of the materials other than water used by energy systems. Several RE systems – notably solar hot water, wind and CST – use more solid materials such as steel and concrete per unit of energy generated than the equivalent fossil-fuelled systems. However, there do not seem to be significant environmental or economic constraints from this use of these bulk materials apart from GHG emissions.[7] For cement, alternatives with low GHG emissions have been developed in the laboratory, but have not yet been commercialised and certified for construction.[8] Biomass could be substituted for coke in steel-making, provided it is obtained in an ecologically sustainable manner.[9] Special attention must be given to a few materials that are currently in short supply, such as the rare earth element neodymium used in the permanent magnets of wind turbine generators. Mark Jacobson from Stanford University and Mark Delucchi from University of California at Davis have jointly researched the options for this and other potentially scarce elements and argue that no major bottlenecks exist in the foreseeable future, taking into account that increasing scarcity will drive increasing prices of these materials which in turn will encourage substitution, the reuse of components after renovation and the recycling of materials that cannot be reused.[10] Others suggest that it would be wise to act now to develop alternative materials and alternatives to the existing sources, which are in some cases only being developed in a single country. Markets are much less effective in managing medium- and long-term issues; the short-term is their main focus. So governments must create incentives to develop alternatives.

## Other environmental, health and social impacts

Here we consider impacts other than energy inputs and associated $CO_2$ emissions, water use and solid materials use, which were discussed earlier.

### Impacts of wind energy

There is irony in the situation that the RE technology facing the fiercest opposition from members of the public is actually one of the most benign in terms of environmental and health impacts: wind power. Unlike fossil fuels, during operation wind turbines emit no

GHG or other air pollutants and use no water. Turbines are almost completely recyclable. They are usually installed on agricultural land or brown-field sites, where their impacts on habitats and ecosystems are small. As discussed above, the energy payback period is much shorter than those of fossil and nuclear power stations. In my view environmentalists are justified in opposing inappropriate siting of particular wind farms that would, for example, involve the clearing of native forests or impacts on wetlands, but not in opposing wind farms in general.

The poorly justified opposition comes mainly from two categories of people: rural dwellers who wish to see a 'natural' rural landscape uncluttered with 'industrial' structures and vested interests who see wind as the principal threat to their favoured conventional energy generation systems, fossil and nuclear. The first category, called NIMBY (Not in My Back Yard) by some people, sees farmland that has previously been stripped of forests as 'natural', even though it has nothing like its original appearance and in many places has been severely damaged by forestry and agriculture leading to depleted soils, erosion, stream banks trampled by stock and loss of biodiversity. Nevertheless, members of this group see wind turbines as local 'visual pollution', while they are unaware of and uninterested in the extensive local damage of coal mining elsewhere and cannot envisage the global destruction being caused by invisible GHGs. While they present themselves as environmentalists, very few members of this group have been active campaigners on any environmental issue prior to or during their opposition to wind farms. Face-to-face conversation reveals that their principal concerns are the visual impact of large wind turbines and fears about loss of the value of their properties; however, they are prepared to campaign on any issue, however tenuous, that could stop wind farms in their 'backyards', namely:

- Bird and bat kills are rare at all but a handful of the tens of thousands of wind farms around the world. They are minimised by careful planning and siting.
- Alleged loss of land value. Most surveys, including a very

large study by the Lawrence Berkeley National Laboratory in the USA,[11] show no loss, while a few show temporary loss immediately after announcement of a wind farm, which disappears over time. However, it could become a self-fulfilling prophecy in a locality where anti-wind campaigners channel their efforts into scare campaigns.

- Fires in the nacelles of wind turbines. These are very rare and have never started a wild fire.
- Noise. This is usually very low at a distance of 500 metres or more. Special cases can be handled by means of strict noise guidelines and monitoring if there are problems.[12]
- The alleged 'wind turbine syndrome' that is supposed to be caused by infrasound, that is, sound that is too low in frequency to be detected by the human ear.

A scientific review of these and other anti-wind-energy claims is given in the Intergovernmental Panel on Climate Change (IPCC) *Special Report on Renewable Energy*. Most impacts are either non-existent or greatly exaggerated.[13]

The last of the above claims has succeeded in instilling fear of wind farms in some Australians, although it is not supported by a shred of scientific evidence. What has been shown is that people in northern Europe, where a high density of wind farms exists, and farmers with wind turbines on their properties do not report a greater prevalence of the alleged symptoms. Furthermore, the level of infrasound emissions from wind turbines is very difficult to detect. At a distance of several hundred metres the emissions are orders of magnitude (that is, factors of 10–1000) below levels that most of us experience from traffic on roads or waves on beaches or air conditioners.[14]

A double-blind experimental study by a PhD candidate in psychological medicine, Fiona Crichton, showed that subjects who were primed to expect symptoms reported symptoms after being exposed to infrasound and also after being exposed to sham infrasound (silence). The control subjects, who were primed to expect no symptoms, reported no symptoms from either infra-sound or sham infrasound. The findings indicate that negative health information supplied by anti-wind-farm campaigners to people living in the vicinity of wind farms has the potential to create symptom expectations, providing a possible pathway for symptoms attributed to operating wind turbines. This is an example of the nocebo effect.[15]

The second category of wind farm opponents comprises:

- supporters of the minerals, fossil fuel and nuclear industries
- electricity utilities that generate mostly from fossil fuels
- big industries that use cheap fossil-fuelled electricity, such as aluminium smelting
- some older electrical engineers who are attached to fossil fuel generation
- the Murdoch press/media generally and
- various organisations that deny climate science.[16]

The two categories of wind farm opponents are linked, in that the second group provides a source of propaganda material, support in the media and possible funding for the first group. An example is the Country Guardian, a group based in the UK, that sends out misinformation (in my view) about wind power to anti-wind NIMBY groups all over the world. Its vice-president, Sir Bernhard Ingham, is secretary of Supporters of Nuclear Energy (SONE), a nuclear power-promoting non-government organisation in the UK. In Australia there appear to be links between the anti-wind-energy group Landscape Guardians and the fossil fuel industry via the so-called Australian Environment Foundation, a front group of the Institute of Public Affairs, which is funded by the mining industries. Despite their name the Landscape Guardians have never campaigned against mining or the logging of old-growth forests, both of which have devastating environmental and land-scape impacts in Australia.[17]

Despite the efforts of anti-wind-farm campaigners, public opinion surveys consistently find wind energy to be widely supported by the general public. However, following announcements of wind farms and vigorous campaigns by opponents, people become confused and local opinion sometimes becomes more evenly divided.

### Impacts of solar energy

The impacts during operation of solar PV are negligible. With well-regulated manufacturing plants, occupational health and safety impacts are low. However, land use deserves a comment.

Although the rate of solar energy capture per square metre of land by PV and CST is very low compared with that of fossil and nuclear energy, it is very high compared with that of agriculture. Agricultural crops need very large areas of land for food production, because the efficiency of conversion of solar energy into stored energy by green leaves is typically a fraction of 1 per cent. This may be compared with the efficiencies of flat-plate solar PV systems of 11–20 per cent for commercial silicon crystal modules, 5–13 per cent for thin film modules and 25 per cent for silicon solar cells in the laboratory. Concentrating collectors have even higher efficiencies. Most residential electricity demand and a significant fraction of commercial demand could be supplied by rooftop solar PV collectors.

At present it does not appear likely that small-scale CST systems will be able to compete with PV on rooftops. Although large-scale power stations require land, it is likely that most will be built on land in regions of high Direct Normal Incidence that is marginal for agriculture, including deserts. Inevitably particular species will need protection in particular locations; however, this is not a general barrier for large-scale solar.

### Impacts of bioenergy

Sourcing and processing biomass feedstocks, particularly dedicated crops, for bioenergy needs great care to avoid adverse environmental and social impacts. The principal environmental issues are GHG emissions resulting from land-use change and from the use of fossil fuels in the life-cycle of the bioenergy source; competition for land between food and fuel; and loss of biodiversity resulting from land clearing for bioenergy plantations.[18] In the ideal case the $CO_2$ emissions from the combustion of biomass are balanced by the absorption of $CO_2$ when the plants that are the source of biomass regrow, with the outcome that the whole process is carbon neutral. However, in practice fertiliser made from oil or gas may be used; the harvested biomass may be transported long distances by trucks fuelled with diesel; and alcohol made from the biomass may even be distilled by burning coal. In the worst examples, such as the production of ethanol from corn in the USA, the life-cycle GHG emissions from the biofuels may sometimes be comparable with those of gasoline and diesel. In Indonesia and Malaysia native forest is being cleared on a large scale to grow palm trees to produce palm oil for the biodiesel market in Europe. The replacement of dense jungle with a low-density plantation generally results in net GHG emissions, a huge loss of biodiversity including orang-utans, and the forced displacement of indigenous people from being self-sufficient hunter-gatherers to sick inhabitants of what are essentially refugee camps.

Environmental and social impacts of biofuel production can be greatly reduced by using residues from agriculture and forestry; using abandoned, degraded or marginal land for dedicated biofuel crops; increasing crop yields and livestock production efficiency to reduce the demand for agricultural land; strengthening the protection of native forests and other natural areas; certifying biofuel production for sustainability; and compensatory off-site mitigation for residual direct and indirect impacts. Ligno-cellulosic feedstocks and algae, still undergoing research and development (R&D), may have much lower impacts than crop-based feedstocks, but it will still be some time before they can be produced on a large scale.[19]

### Impacts of other RE sources

Large reservoirs can have severe environmental and social impacts, whether they are built for hydro or tidal power. Environmental impacts may include loss of biodiversity and, in the case of reservoirs with extensive surface areas covering dense vegetation (as in Brazil), substantial greenhouse gas emissions in the form of methane. Social impacts are the displacement of hundreds of thousands of people, as in the case of the Narmada River scheme in India, possibly even millions in the case of the Three Gorges Dam in China.

Drilling for engineered geothermal power can trigger micro-earthquakes and so is best done in locations far from towns.

So far there is no evidence of adverse impacts of wave or ocean current power.

## Quantified costs and unquantified benefits

For most RE technologies the fuel is free and so almost all the cost is capital cost, which is generally measured in cost per kilowatt of electrical (eg, \$/kW or \$/kW$_e$) or of thermal capacity (eg, \$/kW$_{th}$). The capital cost and the RE resource at the project site are two major determinants of the cost of a unit of energy of a project. Other determinants are the cost of capital, which is measured by the interest or discount rate; the balance of system costs such as inverters and transformers; installation costs; operation and maintenance costs; and, in the case of bioenergy, fuel costs. The cost of energy is simply the sum of all the costs, fixed and variable, over the lifetime of the system, divided by the total energy generation over the lifetime. Actually, the calculation is more complicated because it takes into account that costs incurred or revenue received in the future are discounted. This is done by calculating the levelised cost of energy, *LCOE*, which is the minimum price at which energy must be sold for an energy project to break even (or to have present value of zero). There are various formulas for *LCOE*, a simplified one being:

$$LCOE = \frac{capital\ cost \times R}{(capacity \times 8760 \times capacity\ factor)} + O\&M\ cost \qquad \text{(Equation 5.3)}$$

where O&M denotes operation and maintenance, LCOE and O&M are in dollars per megawatt-hour ($/MWh), capacity is in MW, annual energy generation (the denominator of the fraction) is in MWh, 8760 is the number of hours per year, the capacity factor is a number between 0 and 1 defined in Chapter 1, 'Balancing electricity demand and supply', and R is the *capital recovery factor* defined by

$$R = \frac{i\,(1 + i)^n}{(1 + i)^n - 1} \qquad \text{(Equation 5.4)}$$

where $i$ is the discount rate (see Glossary) expressed as a decimal number, not a percentage, and $n$ is the lifetime of the system in years. Although the formula for R looks complicated, it is really just a modified interest rate; for instance, by inspection it can be seen that in the limit when $n$ becomes very large ($n \to \infty$) and $i$ is fixed, R reduces to $i$ ($R \to i$); also, when $n$ is fixed and $i$ is large $R \to i$.

The numerator of the fraction in Equation 5.3, *capital cost* $\times$ R, is just the annual discounted repayments on the capital cost of the power station; the denominator is just the annual energy generation. As an example, if we choose a wind farm with lifetime $n$ = 30 years and a discount rate of 5 per cent, that is $i$ = 0.05, then Equation 5.4 gives R = 0.065 or 6.5 per cent. Substituting R from Equation 5.4 into Equation 5.3 and choosing a capital cost of $200 million for a capacity of 100 MW, capacity factor 0.3 and an operating cost of $10/MWh (= 1 c/kWh) gives

$$LCOE = \frac{2 \times 10^8 \times 0.065}{100 \times 8760 \times 0.3} + 10 = 51.5$$

Thus, the levelised cost of energy is $51.5/MWh or 5.15 cents per kilowatt-hour (c/kWh).

With the caveat that cost data are sparse and that LCOE

depends on the country, the site within the country and the year in which data were collected, the International Renewable Energy Agency (IRENA) published in 2012 the LCOEs for renewable sources of electricity given in Figure 5.1, assuming a discount rate of 10 per cent per year for all technologies. There are no subsidies included in these estimates.[20]

FIGURE 5.1    Levelised cost of energy estimates for renewable electricity

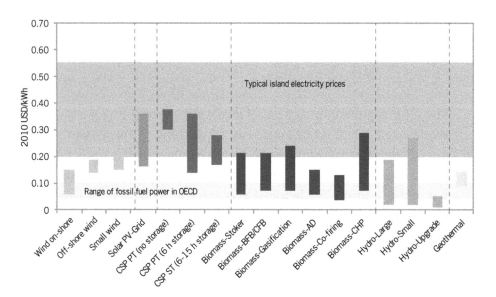

SOURCE IRENA (2012).
NOTES PT = parabolic trough; ST = solar tower; BFB/CFB = bubbling fluidised bed/circulating fluidised bed; AD = anaerobic digester; CHP = combined heat and power; CSP = concentrated solar thermal. The bands reflect typical investment costs (excluding transmission and distribution) and typical capacity factors.

The choice of the same discount rate for all technologies is a simplification; in reality the technologies regarded by financial institutions as more mature would receive lower discount rates than those in early diffusion. Furthermore, technologies with supportive government policies, such as loan guarantees or feed-in tariffs, would be regarded as better risks and so would receive lower discount rates, usually much less than 10 per cent.

Figure 5.1 shows that there is already an overlap in costs between a few RE technologies, notably on-shore wind and some biomass on one hand, and large-scale grid-connected fossil fuel generation on the other hand. It also shows that most of the RE technologies can already compete with conventional island and other remote-area fossil generation, which is usually diesel.

However, comparing the costs of RE technologies with those of grid-connected conventional technologies is complicated, because the principal source of data for the latter is prices rather than costs. Prices depend on variables additional to costs: demand for the technology, which depends to a large degree on government policies, including subsidies; supply (manufacturing capacity), which follows demand after a time delay, while often changing in quite large steps as factories are opened and closed; and the profit made from selling the energy. Furthermore, price comparisons depend on whether the technology feeds energy into the transmission or distribution system or is on the customer side of the meter. For instance, residential rooftop solar PV competes with the retail price of grid electricity, which includes transmission and distribution costs, while a large solar power station competes with wholesale prices. In the latter case, the economic value of a solar power station may also depend on the extent to which it can supply peaks in demand.

For the key RE technologies, the prices have been changing rapidly on an annual basis. The most dramatic changes have occurred with solar PV modules, whose prices decreased steadily pre-2003, then increased slightly until about 2007 as demand for modules exceeded supply, and then fell substantially over 2007–12 as supply overtook and exceeded demand.[21] Bloomberg New Energy Finance estimates that the price of PV modules (presumably some kind of global average) decreased by 80 per cent from 2008 to 2012.[22] Since it is already reaching parity with retail electricity prices for electricity from the grid in several parts of the world, residential PV is now the least dependent on government policies.

Wind turbine prices experienced a similar trend to PV, but with less extreme peaks and troughs.[23] Bloomberg estimates that wind turbine prices fell by 29 per cent on average from 2008 to 2012.[24]

Wind now generates the cheapest electricity in some regions.[25] As the most mature of the new renewable electricity technologies, on-shore wind power is likely to experience only modest price reductions in the future. However, there is still potential for big price reductions in off-shore wind, whose capital costs and operation and maintenance costs are each double those of on-shore wind.

CST power stations, which, like off-shore wind, are in the early diffusion stage of maturity, are still very expensive and have almost ceased to grow in the developed countries hit by the Global Financial Crisis. However, according to engineering studies, they stand poised for price reductions of 30–50 per cent by 2020,[26] provided government policies are implemented to drive the market in a greater number of jurisdictions.

These changes make accurate predictions of future costs and prices uncertain. What we can say is that technological improvements occurring as the result of R&D have still not been translated to commercial products and so, on a timescale measured in decades, all three of the above RE technologies are expected to continue to have declining capital costs, albeit at different rates. On shorter timescales, government policies, discussed in Chapters 8 and 9, must play a vital role in building the markets for RE technologies, as they have and still do for fossil and nuclear power.

Cost comparisons between RE and conventional technologies still have several major shortcomings. The first is that a simple cost comparison assumes that the benefits of both types of technology are identical. This is manifestly false, because RE offers energy security that is – for all intents and purposes – infinite. In addition to providing zero carbon energy supply, RE technologies other than biofuels cause no pollution of air, water and soils during operation. Furthermore, with the exception of biofuels and large-scale hydro and tidal power with dams, RE technologies occupy and damage less land and have less adverse impact on biodiversity than fossil fuels.[27] Because all these benefits of RE are difficult to quantify, they are usually ignored in economic analyses. At best, some countries have a carbon price, but this is usually very small and is not intended to cover all the adverse environmental and health impacts of fossil fuels.

Cost–benefit analysis also fails to account for risks that are rare but devastating in consequences. As discussed in Chapter 6, the spread of nuclear energy helps to increase the number of countries with nuclear weapons and hence the risk of nuclear war. The nuclear disasters at Chernobyl (1986) and Fukushima (2011) are another illustration. If the wind had been blowing from Fukushima towards Tokyo, millions of its inhabitants could have been killed or injured. Yet studies of the external costs of nuclear energy ignore these risks and at best calculate the impacts of radiation releases during normal operation of nuclear power stations and from uranium mines. These are the smallest of all risks on an annualised basis at the margin. However, if we integrate these over the period at which some of the radioisotopes remain dangerous, say 100 000 years, the impacts become substantial, and this brings us to another limitation of cost–benefit analysis as it is practised.

Cost–benefit analysis is generally applied at the margin, an approach of limited usefulness for planning a long-term strategy. It may tell us, for example, that shale gas is currently the cheapest replacement for coal in the USA, but this is of little relevance if we are planning a transition to a zero carbon energy system over the next 20–40 years. We must compare different scenarios spanning such periods. We must include as far as possible all the environmental and health costs of the transition and we must reject technologies that have even small probabilities of events with huge disasters.

## Jobs in sustainable energy

Comparisons of employment in different industries are fraught with conceptual difficulties and inadequate data. Since the definition of 'job' is imprecise, it is preferable to use 'job-years', where one job-year is equivalent to one full-time job for one year. For comparisons to be meaningful, they should compare like with like:

either direct employment or the sum of direct and indirect employment for each industry.[28] For energy technologies one compares job-years per unit of energy generated or saved.

The method giving the most complete picture of employment impacts across the whole economy is Input–Output Analysis. It treats both employment multiplier effects and the impacts of shifts between sectors; that is losses in one sector (eg, coal mining) created by the growth of another sector (eg, wind energy). Thus it handles direct, indirect and induced employment, where the latter comprises for example, the teachers needed to train the workforce for the technology. However, this approach is complex and difficult to interpret. A simpler but less comprehensive method is bottom-up analysis, which counts direct jobs and sometimes indirect jobs.

Researchers at the University of California Berkeley have compared 15 studies on job creation in the USA by a hypothetical expansion in RE, EE and other low-carbon technologies. To do this they first 'normalised' the data in all the studies, that is they converted the data to a form in which they could be compared. They found that all the low-carbon technologies create more job-years per unit of energy than coal or gas. EE in particular does best and moreover is generally least-cost and most easily implemented.[29] Studies from other parts of the world obtain similar results.

Critics of this result argue that creating more jobs per unit of energy is the result of sustainable energy industries being less economically efficient. 'Is it really better for a farmer to hire 50 people to dig up a field than to buy a tractor?', they ask rhetorically. However, those who make use of this simplistic example overlook the fact that very many EE technologies and measures are economically competitive everywhere with energy supply by fossil fuels. Wind energy and solar PV are also becoming competitive in several regions of the world. If subsidies were removed from all energy sources and if external costs were included, it is likely that all commercially available RE sources would be the least-cost options while providing the most jobs. If the economics of a sustainable energy technology are equal to or better than that of the equivalent fossil fuel technology, what does it matter if the former is more labour intensive?

Another subtlety overlooked by the critics is that almost all job comparisons are performed for a particular country or region. While it is possible that the global total of job-years per MWh for (say) a wind farm may be the same as for a coal-fired power station equivalent, the smaller scale of many wind turbine components makes local manufacture feasible. For example, when a coal-fired power station was built in Australia in the early 2000s, only 25 per cent of the capital cost was spent in that country. However, the local content of wind farms was over 50 per cent and job-years created were two to three times that of coal.[30]

Local manufacture, and hence local jobs, will expand as peak oil and an increasing carbon price on transport fuels make their presence felt. Despite the current shake-out in the industry, it may even be possible for manufacturers of solar PV in Germany, the USA and elsewhere to compete with China over the medium term and beyond. The absence of adequate consumer protection laws in China means that there is a large variation in quality between the best and worst solar modules manufactured there. This, together with rising wages in China and increasing transport costs will reduce the cost advantage that Chinese products have at present.

For some low-carbon technologies most potential jobs are in manufacturing (eg, components of wind turbines, CST collectors and railway carriages); others are in installation (eg, insulation, small- and medium-scale solar PV) and still others in both (eg, solar hot water). Some industries could be almost unchanged, apart from additional jobs in R&D: for example, replacing Portland cement with Eco-cement and replacing nitrogenous fertiliser with a climate friendly substitute. Bioenergy can create more employment in rural areas where there are at present very few job opportunities.

Inevitably there are winners and losers in the transition to an ecologically sustainable, energy secure future. Jobs in fossil fuels will decrease, while jobs in EE, RE and public transport will increase. Therefore government policies will be needed for ensuring a socially just transition by funding the retraining and, where necessary, relocation of workers disadvantaged by the transition. Unfortunately some governments, articulating the dogma of 'leave it to the market', appear reluctant to make any effort to smooth industrial transitions.

As many countries still struggle with severe recession, job creation from sustainable energy could assist the recovery while building essential infrastructure for a sustainable future.

## Conclusion

No energy technology is entirely benign. Nevertheless, energy efficiency (EE) and most RE sources that are sited sensibly are everlasting and safe, have very low environmental and health impacts, can create many local jobs and have rapid energy payback periods. In terms of these criteria they are far superior to fossil and nuclear energy. Most of the criticisms of RE technologies are either unfounded or grossly exaggerated. RE technologies that offer some concerns are large dams and some sources of bioenergy.

The economics of energy technologies depend sensitively on the assumptions made:

- whether we include environmental, health and social costs and subsidies in the prices of energy
- whether we place a cost on high risk to present and future generations
- whether we cost a short-term, marginal perspective or a long-term transition to a sustainable energy system and
- whether we choose policies that foster low or high interest/discount rates.

There can be little doubt that stripping away the subsidies to all energy technologies and internalising the externalities would make almost all RE technologies that are currently at the commercial and early diffusion stages of maturity economically competitive with fossil and nuclear energies.[31]

# 6: IS NUCLEAR ENERGY A SOLUTION?

> The nuclear power industry is unintentionally contributing to an increased risk of nuclear war.
>
> Ranger Uranium Environmental Inquiry[1]

## A personal statement

In writing this chapter I'm faced with a dilemma, because many academics, notably scientists and economists, believe that they are neutral and objective and therefore avoid taking up a position for or against a technology or policy. It's part of ensuring their status as respected commentators. However, many studies from the history, philosophy and sociology of science show that scientists have biases, just like other human beings, even when they believe they are being entirely objective. These biases arise from their education and experience, the peers they work with, their sources of funding and the organisations and industries in which they work.[2] If I took the pseudo-objective approach, I would have to give equal weight to pro- and anti-nuclear arguments and pretend to have not drawn critical conclusions. However, to take that approach would be to mislead my readers. It would be like giving equal space to arguments for and against climate science while knowing the claims of deniers have been refuted time and again and that deniers cannot provide an alternative to the human-induced explanation for global climate change that is consistent with the observations. My approach to resolving the dilemma is to state my background, funding and affiliations, to try to be fair and honest in presenting the evidence, and to present my conclusions openly. So, this chapter offers a critique of nuclear energy, giving reasons and referencing the supporting evidence.

Originally educated as a theoretical physicist, my interests spread into interdisciplinary research and teaching of energy and, more broadly, ecologically sustainable and socially just development.[3] My salary is paid by a university. I have never received any payments from the renewable energy (RE) industry, although I occasionally receive research grants from governments for research on RE and energy policy options. Over several decades I have also researched nuclear energy technologies, their impacts and costs. So, here is my take on nuclear energy.

## Introduction

Since 2000 the nuclear industry has mounted a strident international media and lobbying campaign to promote nuclear energy as an alleged 'clean and green' solution to the enhanced greenhouse effect. In the face of all the evidence to the contrary, the industry claims that nuclear energy is safe, emits negligible amounts of $CO_2$, provides 'essential' base-load power and can be rapidly deployed to substitute for coal-fired power stations. The industry further claims that the costs of nuclear energy are generally only slightly above those of coal power and that in some countries nuclear energy is the cheapest form of large-scale electricity.

If these claims were true, then nuclear energy would deserve serious consideration, subject to safety considerations. But, apparently without checking the claims carefully, a number of journalists, including George Monbiot,[4] the environmental scientist James Lovelock and politicians have taken up the nuclear cause. Lovelock, the originator of the Gaia hypothesis, believes that climate change is occurring so quickly that only nuclear energy can provide '$CO_2$-free' energy quickly enough to stop it.[5] He takes the position that, although the environmental and health risks of nuclear energy are large, they are dwarfed by the risks of climate change. This chapter presents evidence that Lovelock and Monbiot were premature in their endorsement of nuclear energy, because they have not addressed some fundamental problems including:

- the environmental and health hazards of nuclear energy, which are much greater than Lovelock and Monbiot envisage, because they ignore its contribution to the proliferation of nuclear weapons and minimise the potential impacts of nuclear accidents involving meltdowns of the reactor core
- $CO_2$ emissions from the nuclear fuel life-cycle
- inherent constraints on the speed of deployment of nuclear energy
- unfavourable economics of nuclear energy compared with those of much cleaner and safer RE sources, such as wind power and solar PV.

Before addressing these fundamental issues, this chapter summarises the current status of nuclear energy, and then sets out the key steps in the life-cycle of nuclear energy, identifying the stages where $CO_2$ is emitted and where nuclear weapons explosives are produced.

## Status of nuclear energy

Nuclear energy arose as a 'spin-off' from nuclear weapons. Its use grew rapidly during the 1960s, nurtured by huge subsidies and the belief that nuclear electricity would soon become 'too cheap to meter'. According to data from the International Atomic Energy Agency (IAEA), in 2012 nuclear power stations generated 2346 terawatt-hours (TWh), 12 per cent less than their peak in 2006. 'About three-quarters of this decline is due to the situation in Japan, but 16 other countries, including the top five nuclear generators, decreased their nuclear generation too.'[6] Meanwhile global electricity demand has been increasing, with the result that nuclear energy's percentage contribution has declined from its peak at 17 per cent in 1993 to about 10 per cent in 2012.[7]

In mid-2013, according to the IAEA database, there were 427 'operating' nuclear power reactors in the world, with total generating capacity of 364 gigawatts (GW). These figures assume the total shutdown of the 10 reactors at Fukushima Daiichi and Fukushima Daini. As a result of the Fukushima Daiichi reactor meltdowns in March 2011, only two of the 44 'operating' reactors in Japan were actually on-line by mid-2013.

Furthermore, in July 2011 the German parliament legislated to shut down all its nuclear power stations by 2022 and commenced that process promptly. In addition Belgium, Switzerland and Taiwan announced that they would phase out nuclear energy over various timeframes. Concerns about hazards and unfavourable economics have stopped the growth of nuclear energy in all but three western countries: Finland, France and the USA. Both the European reactors under construction there are over time by five and four years respectively and 100 per cent over budget (see 'Economics', below). In the USA, no orders for nuclear power stations placed after 1978 have been completed. However, in 2012 the Nuclear Regulatory Commission issued licences to build four new reactors, the first in 30 years, and now two are officially under construction. These two are located in Georgia where the utilities can put the costs into the electricity rate-base, so that electricity customers have to pay for them whatever they end up costing; the projects also receive federal loan guarantees.

There is still significant growth in nuclear energy in China with 28 reactors 'under construction', followed by Russia with nine, then India with seven, South Korea with four and several other countries with one or two.[8] The United Arab Emirates is the first new country in 27 years to have started building a commercial nuclear power plant. China and India are expanding from a tiny base – currently nuclear energy supplies about 2.5 per cent of electricity in China and about 2 per cent in India – and so it will be a long time before nuclear energy has a noticeable impact on the greenhouse gas (GHG) emissions of those countries.[9]

Up to 1989 the number of new reactor start-ups greatly exceeded the number of shutdowns. However, from 1990 to 2012 the reverse was true. It is likely that the Chernobyl disaster in 1986 was the trigger for the switch and that subsequently increasing capital costs took over. The general decline in global nuclear energy may accelerate as many aging reactors are retired over the next 15–20 years.[10]

## Nuclear energy life-cycle

The source of energy in nuclear power stations is known as nuclear fission, the splitting of heavy atomic nuclei into lighter nuclei with the release of energy. An element that can undergo fission, such as uranium-235 (U-235) or plutonium-239 (Pu-239), is called 'fissile'. Fission is initiated by bombarding fissile nuclei with subatomic particles called neutrons. When they split, they release more neutrons, which can split more nuclei and so on in a chain reaction. A chain reaction that grows very rapidly is a nuclear explosion; a nuclear reactor has a controlled chain reaction.

Nuclear power stations use the heat generated from the fission of U-235 and Pu-239 to boil water, to produce steam, to turn a steam turbine, to spin a generator that generates electricity. This may sound simple in theory, but in practice it is a complex multi-stage process.

Figure 6.1 outlines the nuclear fuel life-cycle from the mining and milling of uranium, through the construction and operation of the power station, to the management of spent fuel and the decommissioning of the power station at the end of its operating life. Three of the steps – reprocessing, decommissioning and long-term waste management – are in square brackets, because they are not commonly carried out on a commercial basis, despite several attempts. The original intention was to close the nuclear fuel 'cycle', with unused uranium and newly created plutonium extracted from the spent fuel by reprocessing and fed back into the power station as additional fuel. Since this is not occurring in

FIGURE 6.1    **Nuclear energy life-cycle based on uranium**

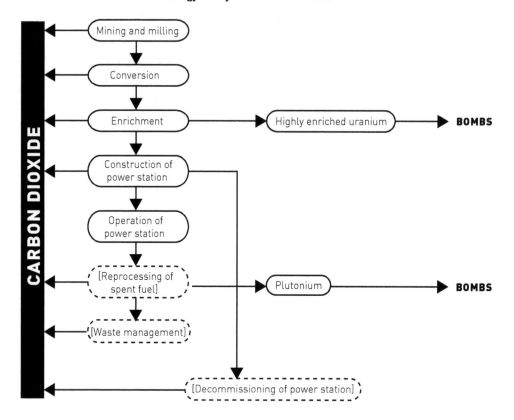

practice to a significant degree, it may be more appropriate to describe the series of processes as a nuclear fuel 'chain' rather than a 'cycle'.

Uranium mining produces uranium oxide, known as 'yellow-cake', and vast quantities of long-lived, low-level radioactive waste that is kept on site. For example, the Olympic Dam copper/uranium mine in South Australia had a radioactive tailings stock-pile of over 136 million tonnes (Mt) at the end of 2012, growing at about 10 Mt per year. It also uses about 12 GL (12 Mt) of artesian water per year.[11]

The element, uranium, exists naturally in two isotopes, which are chemically identical but differ slightly in atomic weight. The common isotope U-238 is not fissile, but the rare isotope U-235 is.

For the most prevalent type of nuclear power station, the light water reactor, uranium is enriched to increase the concentration of U-235 from its naturally occurring level of 0.7 per cent to 3–5 per cent.[12] It is then fabricated into fuel pellets that are inserted into fuel rods and loaded into the reactors.

If uranium is not loaded into the reactor, but rather is further enriched to a much higher concentration, it becomes suitable for use as a nuclear weapons explosive. The nuclear bomb that was exploded over Hiroshima in 1945 used highly enriched uranium.

Since the operation of the nuclear reactor does not involve combustion, there are no $CO_2$ emissions from this stage of the nuclear fuel chain. However, large quantities of cooling water are used. Nuclear power stations in France and the USA have had to be shut down during heatwaves, because the cooling water they were discharging into the lakes and rivers became so hot that it would cause serious environmental impacts.[13] While it is theoretically possible to use air cooling, no single nuclear power station in the world does this. The low thermal efficiency of nuclear reactors would make such a change very expensive.

After a period in the reactor, much of the uranium has been converted into a highly radioactive waste mix. The spent fuel rods are removed and fresh fuel is loaded. Most spent fuel rods are stored for decades under water to allow the radioactive elements with shorter half-lives to decay and to allow the temperature of the spent fuel to drop. Then, as mentioned above, the original idea was to put it through chemical reprocessing, in order to extract the unused U-235 and a new element, plutonium-239 (Pu-239) created by the fission of uranium. Like U-235, Pu-239 is fissile and so it too can be used either as a fuel in a nuclear reactor or as a nuclear weapons explosive. The nuclear bomb that was exploded over Nagasaki in 1945 used Pu-239. If inhaled, plutonium is very carcinogenic – just one millionth of a gram can be deadly.

Because the spent fuel is 'hot', both in terms of radioactivity and temperature, it must be handled remotely, behind shielding. This is difficult, dangerous and very expensive. In the USA three reprocessing plants were built at various times, but none was commercially viable and they have all been closed permanently. The UK reprocessing plant at Sellafield (formerly Windscale) is infamous for numerous discharges into the environment, including the Irish Sea. It was shut down temporarily in April 2005, when it was discovered that 83 000 litres (equivalent to half an Olympic swimming pool) of highly radioactive liquid had been leaking unnoticed for the previous nine months – fortunately none of it escaped off the site. Sellafield reopened in 2007. Reprocessing is still carried out on a large scale at La Hague in France and at small plants in India, Russia and Japan. In practice, very little plutonium is being 'recycled' and vast quantities of nuclear wastes, containing plutonium and highly radioactive fission products such as strontium-90 and caesium-137, are in temporary storage at nuclear power stations (see 'High-level radioactive wastes', below).

## Life-cycle $CO_2$ emissions

The recent push for a revival of nuclear energy has been based on its claimed reduction of $CO_2$ emissions, assuming that it substitutes for coal-fired power stations. In reality, the only $CO_2$-free link in the nuclear fuel chain is reactor operation. All other links – mining, milling, fuel fabrication, enrichment, reactor construction, decommissioning and waste management – use fossil fuels and hence emit $CO_2$, as shown in Figure 6.1.

These emissions have been quantified by researchers who are independent of the nuclear industry. Early work was published by Nigel Mortimer, former Head of the Resources Research Unit at Sheffield Hallam University, UK.[14] Then, in the 2000s, a very detailed study was done by Jan Willem Storm van Leeuwen, a senior consultant in energy systems, together with Philip Smith, a nuclear physicist, both of whom were based in the Netherlands.[15] Since then there have been several additional studies, which have been reviewed by nuclear physicist and nuclear energy supporter, Manfred Lenzen.[16]

Several of the above studies find that the total $CO_2$ emissions depend sensitively on the grade of uranium ore used. Storm van Leeuwen and Smith (SvLS) define 'high-grade' uranium ores to be those with at least 0.1 per cent uranium oxide ($U_3O_8$). In simpler terms, for each tonne of high-grade ore mined, at least 1 kilogram (kg) of yellowcake can be extracted. For high-grade ores, such as most of those being mined in Australia, the energy inputs from uranium mining and milling are small. However, there are significant emissions from the construction and decommissioning of the nuclear power station, with the result that the station must operate for several years to generate its energy inputs. (For comparison, wind power requires only 3–9 months, depending on siting – see Chapter 5.) 'Low-grade' uranium ores contain less than 0.01 per cent yellowcake, that is, they are at least 10 times less concentrated than the high-grade ores. To obtain 1 kg of yellowcake, at least 10 tonnes of low-grade ore has to be mined. This entails a huge increase in the fossil energy required for mining and milling. SvLS find that the fossil energy consumption for these steps in the nuclear fuel chain becomes so large that nuclear energy emits total quantities of $CO_2$ that are comparable with those from an equivalent combined cycle gas-fired power station.

SvLS's work has been critiqued by Lenzen, who argues that a more accurate approach is needed, and then shows that his approach gives much lower values of $CO_2$ emissions from the construction and decommissioning of nuclear power stations.[17] Lenzen also rejects SvLS's requirement that the mine waste should be buried and covered, on the debatable grounds that this recommended safe practice was not carried out for many former mine sites. This value judgement is equivalent to neglecting the risks to future generations of the release of low-level radiation when integrated over 100 000 years.[18]

Table 6.1 compares Lenzen's results for total $CO_2$ emissions from the nuclear fuel chain for a light water reactor with those of SvLS, incorporating Lenzen's corrections to SvLS for construction and decommissioning. Emissions from the construction of large

TABLE 6.1    Total $CO_2$ emissions (g $CO_2$/kWh) from nuclear fuel life-cycle for high-grade and low-grade uranium ore

| U ore grade (% $U_3O_8$) | Emissions excluding mine rehabilitation | | Emissions with rehabilitation | Wind[c] | Natural gas[d] |
|---|---|---|---|---|---|
| | Lenzen[a] | SvLS[b] | SvLS[b] | | |
| 0.15 | 80 | 107 | 117 | 10–20 | 491–577 |
| 0.01 | 131 | 220 | 437 | 10–20 | 491–577 |

SOURCE Lenzen and SvLS, interpreted by Mudd and Diesendorf (2010).[19]
NOTES
(a) Lenzen (2008) excludes significant emissions from the clean-up of mine waste.
(b) I have modified SvLS's results to incorporate Lenzen's corrections for emissions from construction and decommissioning, while keeping SvLS's own results for high-level nuclear waste management. SVLS's results are presented unchanged in Column 4.
(c) Data from Lenzen (2008). See note 16.
(d) Data from ISA (2006). See note 21.

wind turbines[20] are included in the table for comparison, along with those from natural gas-derived electricity.[21]

Comparing Columns 2 and 3 from Table 6.1, it can be concluded that Lenzen confirms SvLS's qualitative result that has been ignored or obscured by nuclear power proponents, namely that there is a big jump in $CO_2$ emissions from the nuclear fuel chain in going from high-grade to low-grade U ores. Furthermore, both Lenzen and SvLS find that, when the uranium ore-grade is low, $CO_2$ emissions from nuclear power are much greater than those from wind power. If we require that mine waste should be covered, but not to the extent of SvLS, then the emissions from the nuclear fuel chain with low-grade uranium would lie between the results of Lenzen and SvLS, at about 300 grams of $CO_2$ per kilowatt-hour ($CO_2$/kWh). These are sufficiently high to provide the basis for the case that nuclear energy, based on existing commercial technology, cannot be a long-term energy/electricity solution to global climate change.

The quantity of known uranium reserves with ore grades richer than the critical level of 0.01 per cent are very limited. Although the quantity is subject to debate, it is generally agreed that the vast majority of the world's known uranium resources are low-grade. With the 2012 contribution by nuclear energy of about 10 per cent of the world's electricity production, the high-grade reserves would at best last several decades. If it were possible to expand nuclear energy rapidly to contribute (say) half of the world's electricity, high-grade reserves may only last a decade or two. No doubt more reserves of high-grade uranium ore will be discovered, perhaps even doubling current reserves, but this would be an insufficient substitute for coal.

Are there alternative future pathways for nuclear energy that could have lower $CO_2$ emissions? Although there are vast quantities of uranium oxide in the Earth's crust, almost all of such reserves exist at very low concentrations, typically $4 \times 10^{-4}$ per cent, at which 1000 tonnes of ore would have to be mined to obtain 4 kg of uranium in the form of yellowcake. In this case the energy inputs to extract uranium would be much greater than the energy outputs of the nuclear power station. Sea-water contains uranium at a concentration of about $2 \times 10^{-7}$ per cent, meaning that one million tonnes of sea-water would have to be processed to extract just 2 kg of uranium. Would new types of nuclear power station resolve the problem?

## Alternative nuclear technologies

### Fast breeder reactors

In theory a fast breeder reactor 'breeds' 50 times more fuel, in the form of Pu-239, than the fuel it consumes. It does this by having a 'blanket' of U-238 around the core of the reactor where the principal reactions take place. Although U-238 is non-fissile, neutrons emerging from the core convert some of it into Pu-239, which is fissile. Thus new nuclear fuel is created without producing additional $CO_2$ emissions.

The world's largest fast breeder reactor, the French Superphénix, was launched in 1974, connected to the grid in 1986 and was closed in 1998 after many technical problems. It only operated for 10 months and cost €12 billion (2010 currency), excluding decommissioning.[22] At present no commercial-scale fast breeders are in operation. A Russian 600 megawatt (MW) demonstration fast neutron reactor, Beloyarsk, is operating, but it has a history of accidents and it is unclear whether it has ever operated as a breeder.[23] The pro-nuclear MIT study of 2003 did not expect that the breeder cycle would come into commercial operation during the following three decades.[24]

Even if fast breeders were to be built in the future, large-scale chemical reprocessing of spent fuel would be necessary to extract the plutonium and unused uranium. Since spent fuel is intensely radioactive, reprocessing has its own hazards and is very expensive to do on a commercial scale.

### Integral fast reactor

The integral fast reactor is a proposed variant of the fast breeder, called 'integral' because it would have an on-site reprocessing plant. The latter would perform a new, experimental kind of reprocessing called pyroprocessing, which would separate the long-lived radioactive elements (transuranics) as a group from the medium-lived radioactive elements (fission products) as a group. Then, in theory, the group of transuranics could be fed back into the reactor as a fuel, 'burning' them up. Thus the only wastes left would comprise the medium-lived fission products, which decay to the background level of radioactivity after several centuries. Since (in theory) plutonium would not be separated from the other transuranics and the combination of transuranics cannot have a nuclear explosion, the proponents claim incorrectly that the integral fast reactor is proliferation-proof.

The flaw in the proponents' argument is that, once the highly radioactive fission products have been separated from the less radioactive transuranics, it would become much easier to extract the Pu-239 from the transuranics and use it to produce nuclear weapons. This appears to be the reason why the US government discontinued research on this system.

## Thorium reactors

Another possible response to the shortage of high-grade uranium ore arises from estimates that there is about three times as much thorium in the Earth's crust as uranium. Although thorium itself is not fissile, it can be converted into an isotope of uranium, U-233, which is fissile, by bombarding it with neutrons. In one approach, the neutrons would be produced by fission of a mixture of U-235 and Pu-239. Then the U-233 would be extracted and fissioned in another reactor. India is attempting to develop this complicated, expensive, two-stage system.

A simpler thorium reactor design would use a particle accelerator to produce the neutrons. This has the advantage that the reactor is fail-safe. Unlike an ordinary uranium reactor, the accelerator-driven thorium reactor can be shut down by simply switching off the particle beam. Furthermore, the nuclear wastes produced by this kind of reactor have much shorter half-lives than from a uranium or plutonium reactor. However, with some difficulty, contrary to the false claims of some thorium reactor proponents, the U-233 could be extracted and purified to make nuclear bombs.[25]

## Nuclear fusion

A possible technology for the long-term future is controlled nuclear fusion, in which the nuclei of light elements, such as the isotopes of hydrogen, namely deuterium ($^2$H) and tritium ($^3$H), are fused to form heavier elements with the release of energy. This is the same type of nuclear reaction that occurs in the interior of our Sun and in hydrogen bomb explosions. Controlled fusion on Earth would have the advantages that there would be much less radioactive waste than from nuclear fission[26] and that one of the fuels, deuterium, would be obtained from a widely available resource, water. However, tritium is rare and must be produced artificially either by breeding it in a fission reactor or bombarding lithium with neutrons.

For half a century, research has continued on the fundamental problem of creating a controlled nuclear fusion reaction and containing it in a laboratory. The challenge is to force the positively charged nuclei together against the force of electrostatic repulsion between them. In the centres of stars, the very hot ionised gas (known as plasma) undergoing the reaction is contained and compressed to high density by the force of gravity. Such high-density plasma cannot be attained in the laboratory, so much energy has to be inputted to produce high magnetic fields to contain the plasma and to heat the plasma to very high temperatures. Plasmas in the laboratory are prone to several types of instability, which terminate their containment. As a result, scientists have been unsuccessful so far in achieving a controlled nuclear fusion reaction in which more usable energy is created than the energy input required to maintain the plasma.

Despite this fundamental shortcoming, the theory suggests that by scaling up from laboratory to large fusion reactor, it may be possible to obtain a net energy gain. But it is also possible that the scale-up will introduce new kinds of instability. Nevertheless, several countries have combined their resources to build an experimental fusion reactor called ITER, a gigantic experiment under construction in France. If all goes to plan, construction will be completed in 2019 and, after years of testing, fusion of deuterium and tritium will commence in 2027. The cost estimate for ITER over its 10-year construction period has escalated and is currently €13 billion.[27]

ITER is not designed to generate electricity – that task would be reserved for the next phase, a prototype nuclear fusion power station called DEMO, that would begin operations in the early 2030s if ITER is successful. Then, if DEMO turns out to be successful during its trial, 'commercial' (that is, at the stage of early diffusion)

fusion reactors might commence feeding electricity into the grid in the 2040s. Thus, by the time fusion is commercially available on a large scale, RE could be supplying most global energy needs, probably at a lower cost.

Proponents claim that fusion could not lead to the proliferation of nuclear weapons. However, in practice, if fusion were ever to become a mature industry, it would probably be coupled with a large nuclear fission industry and thus contribute to proliferation. It could do this producing large quantities of fissile materials (eg, breeding Pu-239 from U-238 or U-233 from thorium)[28] and tritium, an important component of sophisticated nuclear fission and fusion weapons.[29]

### Summary of alternative nuclear technologies

None of the above proposed alternatives to uranium-based nuclear fission – thorium reactor, fast breeder reactor, integral fast reactor or fusion reactor – is commercially available and the latter three are decades away from becoming so. So, on the basis of present nuclear technology and the small existing reserves of high-grade uranium, the potential contribution of nuclear power to the reduction of $CO_2$ emissions is limited.

## Slow deployment

With global warming on a pathway toward 4°C or more, principal technologies for replacing coal-fired power stations and other fossil fuel uses must be implemented rapidly. Nuclear power stations have long planning and construction periods. In the USA Jonathan Koomey from Stanford University and Nathan Hultman from Georgetown University jointly found that the median construction period of US stations for which they had cost data was nine years.[30] Planning and permitting takes additional years. In countries that do not already have nuclear energy, even more years have to be added to build infrastructure and institutions.

For comparison, most improvements in EE discussed in Chapter 4 and several RE systems discussed in Chapters 2 and 5 have very short deployment periods. For example, large wind farms and solar power stations can be planned, approved, constructed and commissioned in two to three years. Some residential solar PV modules are ordered and installed in a few hours. The factors limiting the very rapid deployment of RE are scaling up manufacturing, training the labour-force, building new transmission links and the reluctance of many governments to undertake effective action. Ways of overcoming the latter barrier are discussed in Chapter 11.

## High-level radioactive wastes

After the used fuel rods have cooled for several decades in 'swimming pools' at the reactor sites (see 'Nuclear energy life-cycle', above), the very short-lived isotopes have decayed away and the remaining wastes are stored temporarily above ground where they are vulnerable to terrorist attack. These highly radioactive wastes are of two kinds mixed together: fission products and actinides or transuranics.

Fission products, such as strontium-90 (Sr-90) and caesium-137 (Cs-137), have half-lives of about 30 years and so decay down to natural background levels on timescales of 300–500 years. Actinides or transuranics have much longer half-lives; for instance Pu-239 has a half-life of 24 000 years. Both kinds of high-level wastes are very dangerous to living organisms. If Sr-90 enters the human body, it is stored in the bones where it can induce bone cancer, while Cs-137 goes into the blood where it can induce leukaemia. If inhaled, Pu-239 can induce lung cancer. So the spent fuel must be isolated from humans and other biota for hundreds of thousands of years, even one million years according to some studies.[31]

Not one single repository for the long-term disposal or management of high-level radioactive wastes exists anywhere in the world and very few countries with nuclear energy or nuclear weapons have identified possible sites. Finland and Sweden are the most advanced – they are digging deep geologic repositories. The USA was previously building a waste repository at Yucca Mountain, Nevada, but this was terminated in 2011 after expenditure of $10–$15 billion. Meanwhile, more and more waste is accumulating. The nuclear industry and the governments that support it are imposing the problem on future generations.

A nuclear power station at the end of its life is also a form of high-level nuclear waste. It must be dismantled and its highly radioactive components disposed of safely. Although several small nuclear power stations have been decommissioned at great cost, this has never been done for a full-size (1000 MW) nuclear power station. Costs could be as high as the original capital cost of the power station. In the UK in 2006 the Nuclear Decommissioning Authority (NDA) originally estimated the cost of decommissioning and cleaning up the country's existing nuclear facilities at about £70 billion and subsequently the UK Chancellor Gordon Brown announced that the Treasury increased this estimate to £90 billion.[32] The NDA's current estimate is £59 billion,[33] but this is a 'discounted lifetime cost' to the year 2137 that appears to have been calculated in a different manner from the earlier cost estimates.

## Accidents

The nuclear fuel chain in general, and nuclear power stations and reprocessing plants in particular, are very complex systems that use very dangerous materials. The complexity leads to a high frequency of accidents and the dangerous materials provide the potential for very severe consequences.

The nuclear industry has succeeded in obscuring many accidents by labelling the vast majority as 'incidents' and covering up some of the others. However, over periods of decades information leaks out and several scholars have documented many accidents that have received little or no publicity in the media. A recent compilation by Benjamin Sovacool reveals 99 nuclear accidents from 1952 to early 2010 totalling $20.5 billion in damages worldwide.[34]

Sovacool defines an 'accident' as an event that resulted either in the loss of human life or damages of greater than $50 000.

Of course, cancers do not have little flags identifying whether they were caused by ionising radiation or tobacco smoking or something else. Epidemiological studies following nuclear accidents are constrained by inadequate databases of radiation exposure and of subsequent illness, and the fact that cancer is one of the principal causes of death, even in the absence of nuclear accidents, and so observation of an increase of even (let's say) 100 000 deaths in an exposed population of 100 million is impossible. So it is necessary to make inferences from different events under different conditions, such as the victims of the nuclear bombs on Hiroshima and Nagasaki. Clinical observations can also provide valuable data, but inevitably under-estimate the scope of deaths and illnesses.

The accidents that do succeed in reaching the public eye are the biggest: the disasters at Chernobyl, Ukraine, commencing on 26 April 1986 and at Fukushima Daiichi in Japan, commencing on 11 March 2011. Both were classified as Level 7, the highest level on the International Nuclear Event Scale.

## Chernobyl

At Chernobyl, operator error caused an uncontrolled chain reaction, which triggered a steam explosion that spewed vast quantities of radiation over much of western USSR and Europe. Belatedly about 350 000 people were evacuated from the region. The international nuclear power industry continues to utter the absurd claim that only 31–64 deaths occurred, but these were only the immediate deaths of reactor staff and emergency workers resulting from acute exposure. Estimates of future cancer deaths vary widely. The Chernobyl Forum, a group of UN agencies led by the IAEA, a promoter of nuclear energy, estimated only 4000 deaths.[35] However, the IAEA has the responsibilities of simultaneously discouraging nuclear weapons proliferation while promoting nuclear energy; contradictory goals. Other researchers of the health impacts pointed out that the Chernobyl Forum limited its analysis to local people exposed to quite high radiation doses. Taking into account the much larger numbers of people exposed further afield to lower doses, Elisabeth Cardis from the International Agency for Research on Cancer and colleagues estimated that 16 000 cancer deaths (including leukemia) would be induced in Europe by 2065.[36] A team of 51 medical researchers from Ukraine, Russia and elsewhere contributed new data and estimated that 93 000 cancer deaths plus many other medical conditions may eventually be caused by the accident.[37]

## Fukushima

Before the disaster at Fukushima Daiichi nuclear power station, its owner-operator, the Tokyo Electric Power Company (TEPCO), had ignored warnings of the risk of flooding by tsunamis by both external experts and an internal report. Following the accident, the report of the Nuclear Accident Independent Investigation Commission, chaired by Kiyoshi Kurokawa, former president of the Science Council of Japan, concluded that

> The TEPCO Fukushima Nuclear Power Plant accident was the result of collusion between the government, the regulators and TEPCO, and the lack of governance by said parties. They effectively betrayed the nation's right to be safe from nuclear accidents. Therefore, we conclude that the accident was clearly 'manmade'.[38]

In the event the earthquake shut down the reactors and then the tsunami flooded the diesel generators of the emergency core cooling system. The reactor cores overheated, due to radioactive decay that continued after reactor shutdown. Meltdown followed by hydrogen explosions occurred in three of the six reactors. Meltdown also occurred in one spent fuel storage pool. The local region was severely contaminated and 180 000 people were evacuated.

Contingency plans were made to evacuate Tokyo, but that would have been impossible within the short period of grace that would be given by a change in wind direction. Thus 36 million people were forced to play Russian roulette with a radioactive bullet. Fortunately, much of the highly radioactive material that escaped from the reactors and spent fuel pool drained into the sea or was blown off-shore.

As of August 2013, radiation continues to escape from the crippled reactors into the Pacific Ocean, creating a kind of 'continuous accident' that could span decades.[39]

Although the expected number of cancers is very difficult to estimate, given the absence of comprehensive data on radiation exposure, the other human impacts are clearly profound, with evacuations, families split, widespread psychological illness, industries and livelihoods destroyed, and even fear of and prejudice against the refugees from the region, reminiscent of that against the *hibakusha* (survivors of the nuclear bombings of Hiroshima and Nagasaki).[40] One small positive impact has been the Japanese government's approval of feed-in tariffs for RE.[41] Before then, the political power of the nuclear industry had held back the development of RE in Japan.[42]

## Proliferation and terrorism

Even a regional nuclear war, such as between India and Pakistan, could bring on a 'nuclear winter' resulting in global agricultural collapse for several years and mass starvation,[43] this in addition to the regional devastation from the blast, firestorms and irradiation. There is ample evidence, presented below, that so-called 'peaceful' nuclear power has contributed, and is contributing, to the proliferation of nuclear weapons and so is increasing the risk of future nuclear war. For a start, the education and training of nuclear scientists and engineers for 'peaceful' purposes provides most of the knowledge needed to produce nuclear weapons. Next, the 'peaceful' nuclear life-cycle provides the nuclear explosives, as illustrated in Figure 6.1.

As mentioned in 'Status of nuclear energy', above, there are two principal pathways to explosives for nuclear weapons: uranium enrichment and reprocessing of spent fuel from uranium cycle nuclear reactors to extract the plutonium. If the thorium cycle becomes commercial, an additional pathway would be the separation and purification of U-233 from the inputs to, or the spent fuel from, a thorium reactor.

Uranium enrichment involves processes that take advantage of the slightly different atomic weights of the two isotopes U-235 and U-238. The uranium is first combined with fluorine to form a highly toxic gas, uranium hexafluoride ($UF_6$). This gas was responsible for much illness and even several deaths during the Manhattan Project, the US project to build the first nuclear bombs during World War II. (Incidentally, award-winning investigative journalist Christopher Bryson describes how the image of fluoride was changed from a harmful pollutant into a preventive medication that allegedly reduces tooth decay.)[44] Originally, the uranium isotopes were separated by means of gaseous diffusion, but this required a huge plant and a vast quantity of electricity. The current process, using gas centrifuges, still involves a large plant and substantial electricity generation and so can be monitored by those wanting to stop proliferation. However, the experimental SILEX process, using lasers to enrich uranium, is a much more worrying development. It could potentially enrich uranium in a small plant with relatively low electricity demand and so would be undetectable by satellite surveillance and other methods. In May 2013 SILEX announced that the technology had been successfully demonstrated in the USA.[45]

The same process that enriches uranium to 3–5 per cent U-235 for nuclear power stations can be carried further to achieve much higher levels of enrichment for nuclear weapons. Weapons-grade uranium is defined to be at least 20 per cent U-235 or at least 12 per cent U-233. The official concentration for U-235 nuclear weapons is usually given as 80–95 per cent U-235; however, 20 per cent could be used to make a big crude nuclear weapon. At 20 per cent enrichment, a critical mass of 390 kg would be required and such a bomb would be too large to transport discreetly. However, for enrichment to 94 per cent the critical mass is 25 kg. Unfortunately many research reactors use such highly enriched uranium, so they too must be considered a proliferation risk.[46]

Most plutonium for military purposes is made in special nuclear reactors that produce high-purity *weapons-grade* plutonium. Less than 4 kg of weapons-grade plutonium is sufficient to make a nuclear weapon. Plutonium obtained from non-military nuclear power stations is known as *reactor-grade* plutonium. It can be used to make nuclear weapons, but less 'efficient' ones than weapons-grade plutonium, because the nuclear explosive Pu-239 is 'contaminated' with non-fissile isotopes of plutonium. Nevertheless, less than 10 kg of reactor-grade plutonium will suffice to make a bomb. Despite all the evidence to the contrary, some spokespeople for the nuclear energy industry still attempt to deny this.[47] Yet the IAEA and the US Nuclear Regulatory Commission were set up partly to deal with this very problem. The industry's false claim, that nuclear weapons cannot be made from civil nuclear power, has been refuted for decades by scientific, military and regulatory experts, such as:[48]

- Dr Theodore Taylor, leading US nuclear bomb designer (1976): 'With the spread of peaceful nuclear power, more and more countries have the opportunity to acquire bomb materials.'
- Dr Victor Gilinsky, Commissioner of the US Nuclear Regulatory Commission (1977): 'As far as reactor grade plutonium is concerned, the fact is that it is possible to use this material for nuclear warheads at all levels of technical sophistication ... Whatever we might have once thought, we now know that even simple designs, albeit with some uncertainties in yield, can serve as effective, highly powerful weapons.'
- US Department of Energy (1997): 'All of these grades of plutonium [fuel grade and reactor grade] can be used to make nuclear weapons.'

An ordinary 1000 MW nuclear power station produces about 200 kg of reactor-grade plutonium annually, enough for at least 20 nuclear weapons. Although research reactors produce much less plutonium, commensurate with their smaller sizes, they can still produce enough for a bomb, given enough time. Table 6.2 lists the countries that have used either nuclear power or nuclear research reactors (or both) to assist in placing them on a pathway towards the discreet development of nuclear weapons. In compiling this list official denials have been ignored where there is clear published evidence of proliferation.

TABLE 6.2　Nuclear weapons programs from non-military nuclear technology

| Programs that produced nuclear weapons | Programs discontinued before producing weapons |
| --- | --- |
| UK | Brazil |
| France | Argentina |
| India | Algeria? |
| Pakistan | Australia |
| North Korea | South Korea |
| South Africa (program terminated) | Taiwan (twice) |
| Iran (in progress) | Libya |

SOURCES See text.

Referenced comments on the country programs follow.
- Iran is currently using the enrichment pathway to nuclear weapons. It claims that the enrichment plant is only for 'peaceful nuclear power'.[49]
- One of the sources of plutonium for India's nuclear weapons program is the CIRUS research reactor supplied by Canada. Other possible sources are several CANDU nuclear power stations that are not covered by IAEA safeguards. India describes its nuclear tests as 'peaceful nuclear explosions'.[50]
- Pakistan built a uranium enrichment plant for military purposes with the assistance of plans obtained from the

non-military URENCO uranium enrichment research facility from a metallurgist-engineer, Abdul Qadeer Khan, employed at the plant.[51]

- Libya previously embarked on the pathway to nuclear weapons via centrifuge enrichment, using designs and technology supplied by AQ Khan.[52] However, in 2003 it surrendered its nuclear weapons ambitions and revealed all in return for the lifting of trade sanctions.[53]

- For proliferation by North Korea, see Peter Hayes's papers.[54] The enterprising Dr AQ Khan was apparently involved in this program too.

- In the 1970s, South Africa built a uranium enrichment plant that secretly produced highly enriched uranium for its nuclear weapons program, in addition to low-enriched uranium for its Koeberg nuclear power station and medium-enriched uranium for its research reactor.[55] After producing several nuclear bombs, the government dismantled the program in the early 1990s.

- Rivalry between Brazil and Argentina led both countries to pursue secret nuclear weapons programs based on civil nuclear power. Using technology purchased from West Germany, Brazil commenced in 1978 a military nuclear program that included uranium enrichment well beyond the level required for nuclear electricity.[56]

- Argentina pursued its covert nuclear weapons program from 1978, using technology supplied by several countries. It built a uranium enrichment plant and also commenced construction of reprocessing facilities to extract plutonium.[57] In 1992, Argentina and Brazil signed a bilateral agreement on mutual supervision of nuclear facilities and, in 1994, the Tlatelolco Treaty for the Prohibition of Nuclear Weapons in Latin America. Subsequently, both signed the Treaty on the Non-Proliferation of Nuclear Weapons (or Non-Proliferation Treaty – NPT).

- In the UK the Magnox reactors at Calder Hall and Chapelcross were used simultaneously for electricity generation and plutonium production for nuclear weapons, supplementing the plutonium from military reactors.[58]

- In France, there has been a close relationship between civil and military nuclear programs. The Celestin reactors were used for both civil isotope and military plutonium production. Phénix was intended as a prototype for the larger fast breeder power reactor, Superphénix, but its plutonium production appears to have been primarily for military purposes. Statements by French generals indicate that Superphénix, which was shut down in 1998, was intended for both electricity generation and large-scale plutonium production for weapons.[59]

- South Korea conducted secret small-scale experiments on plutonium extraction in 1982 and uranium enrichment in 2000.[60] US pressure stopped the delivery of a reprocessing plant from France to South Korea. South Korea still has a 30 MW research reactor that can produce weapons-grade plutonium.

- Taiwan, which has six nuclear power stations and a CIRUS research reactor, had covert nuclear weapons programs until 1988 at least.[61]

- Australia commenced the development of a nuclear power station at Jervis Bay, which was intended for joint production of electricity and plutonium for nuclear weapons. It was terminated at an early stage by a change of prime minister.[62]

- In the past Algeria was suspected of conducting a nuclear weapons program at its secret nuclear facilities, but the evidence is not as clear as in the other countries listed in Table 6.2.[63]

Thus there is ample evidence of the intimate links between civil nuclear reactors (both for research and electric power) on one hand and nuclear weapons on the other. James Hansen, James Lovelock, George Monbiot and others who support nuclear energy mistakenly assume that the major hazard from nuclear energy is a serious accident at a nuclear power station or reprocessing plant, from which the casualties might range up to several million if a large city such as Tokyo were to be exposed to major fallout. Nuclear accidents can be bad enough, but the outcome of the use of nuclear weapons obtained from nuclear energy could be hundreds of millions of deaths, even billions, taking into account the likelihood that the soot from a nuclear war could cause a nuclear winter that could stop global agriculture for several years.[64] Thus the expected numbers of human casualties of nuclear war could be in the same order of magnitude of those occurring over a long period of time from global climate change from the enhanced greenhouse effect. Ironically a nuclear winter would temporarily halt global warming.

Terrorist attacks on nuclear facilities have become much more likely since 11 September 2001. Although the nuclear industry claims that the containment vessel of a nuclear power station would withstand the impact of a fully fuelled jumbo jet, this has never been tested. More feasible scenarios involve sudden take-overs of nuclear power stations by paramilitary groups, which could initiate a core meltdown from the control room or possibly breach the containment vessel by setting off explosives inside. The transportation of high-level nuclear wastes, by land or sea, is particularly vulnerable to terrorist attack. There is evidence of a black market trade in highly enriched uranium and Pu-239.[65]

## Economics

### Scarce, unreliable data and misleading analyses

A report to the UK Sustainable Development Commission points out difficulties of obtaining objective data on the economics of nuclear power:

> There are few sources of data on the costs of future nuclear
> power that relate directly to UK circumstances … The
> problematic category is capital costs, where there is no recent
> European or North American experience. Examination of the
> limited number of published capital cost estimates that apply

directly to the UK shows that all appear to derive from studies originally designed to apply to other countries *and from vendors of reactor systems*.[66] (my italics)

It is risky to accept manufacturers' and developers' estimates of capital costs and to sign a contract that does not specify a fixed cost, yet that is what some purchasers do.

Furthermore, proponents often mislead or confuse potential purchasers of nuclear power stations and other decision-makers in the following ways:

- Presenting only the 'overnight' capital cost and ignoring the interest during construction, which can be a large fraction of the capital costs, especially when construction is slow.
- Assuming an unrealistically low discount rate to convert capital cost in dollars per kilowatt into a levelised cost of electricity in cents per kilowatt-hour (c/kWh).
- Giving optimistic cost estimates for new or modified types of reactors that are only in the design stage and have not been built. These are almost always under-estimates.
- Quoting a cost estimate for purchase of several reactors when only one is sought.
- Making over-optimistic assumptions about the operational performance of the nuclear power station.
- Omitting to specify the year of the currency.
- Ignoring the subsidies to nuclear energy that continue to be handed out after 50 years of experience.

### Subsidies to nuclear energy

Comprehensive quantitative data on subsidies are incomplete and difficult to obtain. The available data show that subsidies vary in quantity and type from country to country. The existence of subsidies entail that risk is not properly allocated in the market thus masking the true economics of nuclear energy.[67] Reported subsidies include:

- government funding for research and development, uranium enrichment, decommissioning and waste management
- loan guarantees backed by governments (that is, taxpayers)
- stranded assets paid for by ratepayers and taxpayers and
- limited liabilities for accidents paid for by the victims and taxpayers.

In the USA, one estimate of subsidies accumulated over the 50-year period 1948 to 1998 was about $74 billion[68] or around $130 billion in 2012 currency. Another report found subsidies to US nuclear power to be about $9 billion per year in 2006.[69]

The subsidies continue. According to nuclear analysts Mycle Schneider and Antony Froggatt:

> The July 2005 US Energy Policy Act was aimed at stimulating investment in new nuclear power plants. Measures include a tax credit on electricity generation, a loan guarantee of up to 80 percent of debt (not including equity) or $18.5 billion for the first 6 GW (in exchange for a credit subsidy fee that the utility must pay to the federal government and that is calculated based on the borrower's risk of default), additional support in case of significant construction delays for up to six reactors, and the extension of limited liability (the Price Anderson Act) until 2025.[70]

In Germany, a recent study commissioned by Greenpeace found that total (direct plus indirect) subsidies from 1950 to 2008 amounted to €165 billion (US$235 billion).[71] Further discussion on subsidies within the European Union is on the WISE/NIRS website.[72]

When the UK electricity industry was privatised, the British government had to impose a levy on electricity prices, called the Fossil Fuel Levy, to subsidise nuclear electricity by means of the Non-Fossil Fuel Obligation (NFFO). In the 1990s this subsidy peaked at £1.3 billion per year,[73] equivalent to a subsidy of 3 p/kWh, making the total cost of nuclear power at that time about 6 p/kWh.

The British government's commitment to pay the decommissioning and clean-up costs of existing nuclear power stations adds £59 billion (discounted to 2137) to the taxpayers' burden. Back in 2003, the British White Paper on Energy stated that 'the current economics of nuclear power make it an unattractive option for new generating capacity'.[74] However, the UK's electricity generation system is now mostly owned and controlled by French and Germany utilities – EDF, E.ON and RWE – some of which have large involvements in nuclear energy. So it is possible that their influence is responsible for the recent change in the UK government's position towards acceptance of more nuclear power stations. In 2013 the coalition government is negotiating with nuclear suppliers, offering a loan guarantee and a guaranteed fixed price for electricity from a proposed new nuclear power station, while telling the public that there will be no subsidy.[75] Such proposed contracts are subsidies and are likely to be big ones.

While subsidies to RE do exist, they are generally much smaller than nuclear subsidies (see Ch 9, 'Removing subsidies to fossil fuels and nuclear energy') and, unlike the latter, are being reduced rapidly as markets for RE technologies grow and prices decrease.

### Nuclear costs in the USA

Despite huge subsidies, the USA has not completed a new nuclear power station for over 30 years. This has been attributed primarily to poor economics,[76] although the accident at Three Mile Island in 1979 and the anti-nuclear movement have also contributed.

Estimates of nuclear energy costs in the USA have escalated rapidly through the 2000s. Some of these results are summarised in Table 6.3 and Figure 6.2. The latter shows a clear trend of overnight capital cost escalation in real terms among nuclear power stations through the 1980s and into the 1990s, while the studies listed in Table 6.3 suggest rapid cost escalation during the 2000s. Energy analyst Amory Lovins attributes these escalations to 'severe manufacturing bottlenecks and scarcities of critical engineering, construction, and management skills that have decayed during the industry's long order lull'.[77]

TABLE 6.3    Escalation of estimated nuclear power capital costs, USA, 2003–10

| Study | Capital cost (US$/kW) |
|---|---|
| MIT (2003) | 2000 + interest during construction (IDC) |
| Keystone Center (2007) | 3600–4000 |
| Harding (2007) | 4300–4550 |
| MIT (2009) update | 4000 + IDC |
| Moody's (2008) | 7500 |
| Florida Power & Light (2009) | 5455–8180 |
| Severance (2009) | 7400 with no further escalation |
| | 10 500 assuming current escalation rate continues |

SOURCE The author's selection of published studies.[78]

For comparison, quotes for new reactors in Canada, Turkey, United Arab Emirates and Finland were around $11 000/kW in 2009 and 2010.[79]

As shown in Figure 6.2, overnight cost estimates by Wall St and independent analysts range from $6000 to $10 500/kW, much higher than those of early consultants and utilities. Actual capital costs, including interest during construction (IDC) and cost escalation, will be even higher. Assuming conservatively that these additional costs add 20 per cent to overnight costs[80] lifts this capital cost range to $7200–12 600/kW. Keeping the same assumptions as in Chapter 2, 'Maturity of energy technologies' gives a cost of energy range of 11.0–19.2 c/kWh.

In recent years, operating costs in the USA have been quite low, around 2 c/kWh, but this is partly because high capacity factors have been finally achieved after decades of poor performance and partly because the government assumes responsibility for the disposal of spent fuel for the nominal fee of 0.1 c/kWh.[81]

California produces its own analysis of the economics of electricity generation from various technologies. Table 6.4 shows

FIGURE 6.2 **'Overnight' capital costs for existing and proposed US nuclear power stations**

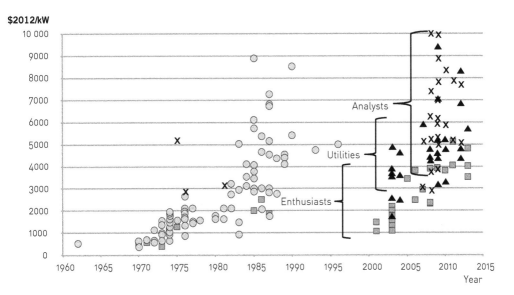

Actual  ■ Enthusiasts (Gov't, Acad, Vendors)  ▲ Utilities  ✗ Analysts (Wall St. & others)

SOURCE Cooper (2009b)[82] reviewing numerous studies.
NOTE 'Overnight' cost is capital cost ignoring IDC.

levelised costs of energy projected in 2010 for 2018 by the California Energy Commission, distinguishing between merchant purchases, investor owned utilities (IOU) and publicly owned utilities (POU), which have different discount rates and other conditions.[83] Projected levelised costs of nuclear energy range from 17 to 34 c/kWh. Since the report was published, solar PV costs have declined substantially.

## Nuclear economics in western Europe

There are only two new nuclear power stations under construction in western Europe. Both are Areva's so-called Generation III+ reactor, featuring small changes in design from Generation II reactors to include some 'passive' safety features. But the reactors

TABLE 6.4 **Projected average levelised costs of electricity generation, California, 2018**

| Technology | Rated power (MW) | Merchant (c/kWh) | IOU (c/kWh) | POU (c/kWh) |
|---|---|---|---|---|
| Gas turbine, open cycle, peak load | 100 | 95 | 74 | 37 |
| Combined cycle gas | 500 | 17 | 16 | 15 |
| Coal IGCC without carbon capture and storage (CCS) | 300 | 18 | 14 | 11 |
| Biomass IGCC without CCS | 30 | 17 | 16 | 14 |
| Nuclear, Westinghouse AP-1000 | 960 | 34 | 27 | 17 |
| Hydro, small, developed site | 15 | 16 | 16 | 12 |
| Solar, parabolic trough | 250 | 30 | 29 | 26 |
| Solar PV, single axis | 25 | 31 | 30 | 26 |
| Wind, on-shore, class 3/4 site | 50 | 13 | 12 | 9 |
| Wind, on-shore, class 5 site | 100 | 11 | 11 | 8 |
| Wind, off-shore, class 5 site | 350 | 211 | 20 | 15 |

SOURCE California Energy Commission, (Klein 2010, Table 5).[84]
Notes: In-service year: 2018; currency: nominal US 2018 dollars. The costs include insurance costs and US federal and Californian tax benefits.

are still not fail-safe in terms of avoiding meltdown accidents. Both of these construction projects are experiencing long delays and hence much higher than expected capital costs. In Finland, the Olkiluoto-3 nuclear power station, 1600 MW, has been under construction since 2005.

Schneider and Froggatt state: 'By early 2012, following a long series of management problems, quality control issues, component failures and design difficulties, Olkiluoto 3 is about five years behind planning and cost estimates rose to between €6 and €6.6 billion or 100–120 percent over budget.'[85]

The nuclear industry originally claimed that the commencement of this project demonstrated that nuclear energy is competitive under market conditions. In rebuttal, independent commentators pointed out that the station was being built by a consortium that includes a 40 per cent share by the government of Finland and that it would sell its electricity to members of its own consortium. Therefore, if completed, it would not operate under conditions of a competitive market and so it can obtain finance at interest rates far below market rates.

In France, construction commenced in 2007 on a new 1630 MW nuclear power station for EDF at Flamanville. Following delays and cost overruns, in July 2011 EDF announced that the estimated capital cost had escalated from €3.3 billion to €6 billion and the completion date had been delayed to 2016.[86]

### Economic implications of Fukushima

The Fukushima nuclear disaster, which commenced in March 2011 and continues in 2013, reinforced the experience gained from the Chernobyl disaster that the costs of rare but catastrophic nuclear accidents are enormous and must be properly insured. The direct costs of the Fukushima event include the costs of housing the 180 000 evacuees from the vicinity of the reactor; damage to local fisheries, agriculture and other businesses contaminated by radioactive elements; clean-up of radioactive land and infrastructure; and the cost of land in the permanent exclusion zone, currently 800 square kilometres in area. In addition, there are health and social security costs, which are difficult to quantify. The Japan Center for Economic Research estimated that partial costs are in the range US$71–$250 billion. The figures include $54 billion to buy up all land within 20 kilometres of the plant, $8 billion for compensation payments to local residents, and $9–$188 billion to scrap the plant's reactors.[87]

The accident caused blackouts over a much wider region and impacted severely on the Japanese stock market and economy. Manufacturers closed their factories and the tourist industry suffered. The costs of these additional impacts are still to be evaluated.

In Japan and around the world, the operators of nuclear power stations are only required to insure for relatively small amounts. TEPCO, the company operating Fukushima Daiichi, was only required to cover about US\$1.5 billion of liability protection for the costs of a radiation leak, but this excludes earthquake and tsunami. Most costs above this are uninsured and must be covered by the government (ie, taxpayers), electricity ratepayers and the victims.

## Conclusion

New nuclear power cannot make an effective contribution to cutting GHG emissions because it is too dangerous, too expensive, too slow and, in the long run, will become too greenhouse intensive. While it is possible that alternative nuclear power technologies that are less greenhouse intensive may enter the market in the long-term future, they would all be still too dangerous in terms of proliferation, even more expensive than the current generation and still too slow. I expand a little here on some of these points.

Since the 1970s experts in nuclear technology have recognised and stated publicly that nuclear weapons explosives – highly enriched U-235 and reactor grade Pu-239 – can be produced from the 'peaceful' nuclear fuel chain, specifically from uranium enrichment and by reprocessing the spent fuel from both power and research reactors. A nuclear weapons explosive, U-233, can also be purified from the thorium fuel cycle. Thus the spread of nuclear energy drives the spread of nuclear weapons capability under the cloak of electricity generation and hence increases the risk of nuclear war. The persistence of nuclear industry associations and other nuclear energy supporters in denying these well-established facts raises questions about their credibility on other nuclear issues, such as the $CO_2$ emissions from the nuclear fuel chain and nuclear economics.

The planning and construction of a nuclear power station takes typically over nine years, much longer in countries that do not have any industry infrastructure apart from uranium mines. Therefore, nuclear energy cannot make a significant contribution to the reduction of their $CO_2$ emissions over the next 20 years. Even in the unlikely event of a future slow expansion of the global nuclear industry, a global shortage of high-grade uranium ore will become a barrier to the industry within several decades. Although there are vast global resources of low-grade uranium ore, production of nuclear fuel from low-grade uranium ore produces quite high $CO_2$ emissions. Therefore, new nuclear power stations based on existing technologies could not substitute significantly for $CO_2$ emissions from coal-fired power stations. They are neither a short-term nor a long-term solution to global climate change.

There are still no permanent repositories for high-level nuclear wastes, although underground repositories are being constructed in Finland and Sweden. Furthermore, the current generation of nuclear power stations requires uranium mining that produces huge mountains of low-level radioactive wastes that will blow harmful dust in the wind for 100 000 years or more. Most nuclear power stations use vast quantities of water for cooling. They expose large populations to the risk of rare but potentially devastating accidents. Their electricity is currently more expensive than wind energy and soon will be more expensive than solar PV in many countries.

The risks of proliferation, terrorism and accidents, taken together with the lack of long-term waste management repositories and the environmental impacts (including $CO_2$ emissions) and high costs of the nuclear fuel life-cycle, characterise a source of electricity that is not, by any reasonable criterion, ecologically sustainable or commercially viable. Thus, spending valuable financial resources on nuclear power would be an expensive diversion from fast, effective, less expensive technologies such as energy efficiency (EE), other forms of demand reduction and RE.

# 7: TRANSPORT AND URBAN FORM

Resilience in our personal lives is about lasting, about making it through crises, about inner strength and strong physical constitution ... Resilience can be applied to cities.

Peter Newman, Timothy Beatley and Heather Boyer[1]

To transform our energy demand and supply into a very low-carbon system, we must transform our cities, especially the urban transport system that forms a similar function to the blood circulatory system in our bodies. This will be a much more challenging task than replacing fossil-fuelled power stations with renewable energy (RE) and energy efficiency (EE). Many people will not experience much difference when they receive 'green electrons' instead of 'black'. Although there will need to be some behavioural changes in reducing energy use in the home and office, much of the task will be taken up by technologies, such as insulation, more efficient buildings and appliances, and 'smart' devices that automatically switch off energy-using appliances and equipment when demand is high and supply is low.

Urban infrastructure is much more pervasive and longer-lived than appliances and even power stations. In a sustainable future people who live in cities and suburbs will be faced with frequent decisions on how to travel from A to B and, given alternative options, whether to travel at all. Gradually urban form will change, becoming much more closely integrated with a growing urban public transport (transit) system and also a growing active transport system based on cycling and walking.

These changes will be driven not only by the need to cut greenhouse gas (GHG) emissions and local air pollution. Car-dependent cities are very vulnerable to interruptions to the supply of oil. In these cities access to food, medical and hospital care, workplaces, educational institutions, and social and cultural activities all depend on a continuous input of oil. Food production in rural areas also depends on oil to make fertiliser and run tractors. Yet, as discussed in the Introduction to this book, the world is close to the peak in oil production and may even have passed it. Many countries are highly dependent upon imports of oil from the finite, decreasing reserves of the few remaining big oil exporters. To make matters worse, many cities in rapidly developing countries are jammed with motor vehicles for most of the day, seven days per week. They are polluted, unhealthy and unable to function properly, economically and socially.

The solution to these problems is obvious, but not easy to implement in existing traffic-clogged cities. It is, in the words of urban transport expert Peter Newman, to make our cities more resilient through the expansion of public transport and active transport, the increase in urban population density around public transport nodes, and the resulting reduction in trips by car.[2] There are also minor roles for improving motor vehicle technologies to make them more energy efficient and for changing fuels from fossil liquids and gases to renewable liquids, gases and electricity. But without radical changes to the whole system of transport and urban form, reductions in emissions will be modest and many cities will face, indeed already face, the danger of collapse.

This chapter focuses on *urban* transport, because more than half the world's population lives in cities and this proportion is growing rapidly, driven by government policies in some countries such as China and by the poverty and lack of opportunities in many rural areas of the world. Furthermore, in many countries the largest quantities of GHG emissions from transport are emitted in cities.

The section entitled 'Impacts of car and truck dependency' below expands upon the adverse environmental, social and economic impacts of car dependency in cities. It is followed by 'Integrated urban and transport planning', the heart of the chapter, which expounds the major solutions of integrated urban and transport planning flagged above; addresses the role of urban density on car use; recommends restructuring cities into population clusters based on access to public and active transport; examines various trip generators; and raises the new issue of the social psychology of choosing modes of transport in cities. The section 'Improved vehicles and fuels' looks at some traditional approaches to the transport problem that offer minor contributions to the solutions: improved vehicle technologies and improved fuels. Finally a short section is dedicated to intercity and rural transport that offers some grounds for optimism based on current trends.

## Impacts of car and truck dependency

The motor car and the truck have made substantial contributions to human civilisation. However, the huge population of motor vehicles and the land and infrastructure they demand in the developed and rapidly developing regions of the world are causing severe adverse environmental, health, social and economic impacts.

### Environmental and health impacts

First let's consider some statistics. In 2010, 43 per cent of global $CO_2$ emissions from fuel combustion were produced from coal, 36 per cent from oil and 20 per cent from gas. Currently the vast majority of transportation is fuelled by oil – very little comes from gas or coal-fired electricity. Transport produced 22 per cent of global $CO_2$ emissions, and of this, road transport was responsible for nearly three-quarters.[3]

In addition to being substantial emitters of GHGs, motor vehicles are the major source of local air pollution in large cities, for example:[4]

- Motor vehicles emit volatile organic compounds (VOCs), which are carcinogenic and toxic.
- Motor vehicles are responsible for the majority of the emissions of carbon monoxide, a toxic chemical.
- Motor vehicles are the main source of oxides of nitrogen (NOx), which are toxic to plants and animals. NOx also react with hydrocarbons to form ozone and photochemical smog, which is responsible for respiratory diseases.
- Sulphur oxides (SOx) are toxic to plants and animals and cause acid rain.
- Diesel-fuelled vehicles emit fine particles, which are toxic and responsible for respiratory diseases and cancer. With Europe in the lead, improved fuel quality standards are reducing these emissions in several countries.
- For people living or working near busy roads, traffic noise can cause stress, insomnia and inability to concentrate.
- Car dependence results in lack of exercise and hence obesity and heart disease.
- Although fuel quality standards are being increased in stages, these improvements are offset to some extent by the increasing number of kilometres driven.

Roads and parking areas occupy and degrade much land in cities, including bushland, market gardens and other agricultural land. In urban centres and subcentres over half the land area may be devoted to motor vehicles; in suburbia typically 20 to 30 per cent of land is tied up in this way. By destroying much habitat, these sealed areas impose huge damage on biological diversity. With run-off from roads and car parks containing oil, grease and heavy metals, motor vehicles damage water quality by polluting creeks, rivers, harbours and beaches.

Roads are responsible for many deaths and disabilities from crashes (see Table 7.1). The rate of injuries is many times that of the rate of fatalities. Economic impacts are huge. Trucks in particular are responsible for a disproportionately large proportion of road deaths and disabilities.

TABLE 7.1    **Annual road fatalities by selected country**

| Country | Total deaths | Deaths per $10^5$ population | Deaths per $10^8$ VKT | Year | Reference[5] |
|---|---|---|---|---|---|
| Australia | 1310 | 5.8 | 0.57 | 2012 | BITRE (2013) |
| China (official, police data) | 81 649 | 6.2 | ? | 2007 | China Autoweb (2013) |
| China (death certificates) | 221 135 | 16.7 | ? | 2007 | China Autoweb (2013) |
| Japan | 4411 | 3.5 | ? | 2012 | Japan Times (2013) |
| USA | 32 367 | 10.4 | 0.68 | 2011 | NHTSA (2011) |

NOTE Data quality from developing countries is generally poor, as demonstrated by the two different estimates for China. VKT is vehicle-km-travelled.

## Economic impacts

A wide range of adverse economic impacts result from society's high dependence on motor vehicles. Car-dependent cities tend to sprawl over vast areas and so are responsible for big additional infrastructure costs compared with compact cities with excellent public transport systems. In sprawling cities, roads, water, sewerage, gas, electricity, telephone, ambulance, fire brigade, schools and hospitals all cost more. Proponents of this kind of urban structure have claimed that car-dependent cities are wealthier and by implication that cars are the cause of this alleged economic benefit. This notion has been challenged by urban transport experts Peter Newman and Jeffrey Kenworthy, who compared car use, measured by kilometres travelled per person per year, with wealth, measured by annual gross regional product (GRP) per capita, in 37 cities grouped into regions in 1990. Although there are no strong correlations, it is clear that cities with high wealth per capita (mostly European and wealthy Asian cities) are associated with lower car use than those in the mid-wealth range (US and Australian cities).[6]

A consulting report commissioned by the Australian Department of the Environment pointed out that:

> Unlike other utilities, roads have not so far been treated as a
> capital asset that should be required to earn a rate of return.
> This treatment would recognise not only the capital value of
> bridges and road pavements, but [also] of the land devoted to
> roads ... The theory of land valuation is that all land should
> be valued at opportunity cost. Applying this to road land
> would result in its site value being inferred from the adjacent
> properties.[7]

The large areas of land occupied by roads and parking have a high value in city centres and subcentres. As a case study, by taking an average real estate value for urban land and a discount rate of 10 per cent real, my former colleagues at the University of Technology Sydney and I found that public land occupied by roads and parking in the Sydney metropolitan area had an annualised value of about $3.5 billion (2001 Australian dollars) in 1996. These results completely change the economics of the various transport modes, boosting the cost of car travel to 72 cents per passenger-kilometre (c/pkm), trains to 45 c/pkm and buses to 29 c/pkm. Thus cars are by far the most expensive urban passenger transport mode, followed by trains and then buses (which only use a small fraction of the road and parking land) as the cheapest. Motorists are unaware of this, because they use the most heavily subsidised mode of transport in Sydney.[8]

Land values are not the only 'external' cost of motor vehicles that have to be taken into account. The cost of traffic congestion is the cost of time losses experienced by other road users as a consequence of a motorist's decision to drive. Again using Sydney as an example, the costs of congestion in Sydney have been estimated to be $12.1 billion in 2005 and projected to rise to $16.6 billion in 2020 under business-as-usual.[9] The same study estimated the cost of crashes in Sydney in 2005 to be $3.9 billion, ill-health from air pollution to be $1.2 billion, subsidies to the Roads and Traffic Authority to be $0.7 billion and GHG emissions (at the very low assumed cost of $10 per tonne of $CO_2$ emitted) to be $0.145 billion. These Sydney figures could be roughly scaled up to the whole of Australia by multiplying by four. Road works are costing Australia over $7 billion per year and the major part of this is caused by heavy trucks.[10] Other subsidies to roads and motor vehicles include tax refunds for motor vehicle use, support to the automotive industry, company tax deductions for petroleum exploration and various grants.[11]

### Social impacts

An automobile-dependent city is one that relies on cars to the virtual exclusion of environmentally sounder modes, such as public transport, cycling and walking. One of the characteristics of such a city is its low population density and its tendency to sprawl over a large area. In Australia an extreme example is the capital city, Canberra, with a population of 300 000, an area comparable with that of Greater London (population 10 million) and a population density of only 10 people per hectare. The government of the Australian Capital Territory, in which Canberra is located, has gained limited control over the motor car by removing free parking in Canberra's city centre and subcentres and providing a good bus service, at least on week-days, considering the city's ultra-low density and large area.

One of the social consequences of urban sprawl is the isolation of low-income earners in outer suburbs with few facilities. Of particular concern are car-less parents caring for young children, and others who do not drive, because they are too old, too young or suffering from disabilities.

In the inner suburbs of automobile-dependent cities, major roads divide communities and produce a hazardous environment for young children, so dangerous that nowadays many parents drive their children to school. Roads, traffic and parking take over public open spaces, either directly or by making them unpleasant through noise and air pollution.

Sprawling car-dependent cities are also very vulnerable to socio-economic collapse if their oil supply is interrupted.

### Summary

Motor vehicles and their infrastructure have grown to the extent that they are damaging environment, health, society and the economy. A valuable servant has been transformed into an almost unconstrained master, especially in cities. It is time to bring it under control.

## Integrated urban and transport planning

A range of social, economic and cultural factors influence people's choice of public or private transport powered by human or fossil fuel energy. Evidence suggests, however, that the most important factor is urban planning.[12] When the arrangement of public and private buildings and other facilities makes human-powered transport and public transport convenient and cost-efficient, people use it. Conversely, when cleaner alternatives are inaccessible, private cars are used regardless of cost. A key factor in urban planning is population density or, more precisely, the geographic distribution of population in relation to public transport nodes.

### Urban density and fuel consumption

In the outer suburbs of most US and Australian cities, the population density, job concentration and average income are low, and reliance on private cars is high.[13] In these areas public transport

FIGURE 7.1 **Petrol energy use in private passenger travel versus urban population density**

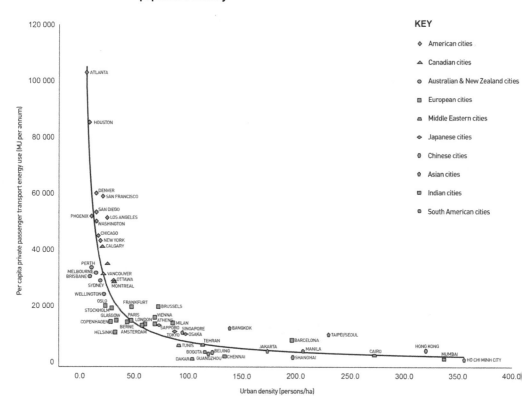

SOURCE Newman and Kenworthy, Global Cities Database, personal communication, updated from earlier data in Newman and Kenworthy (1999).

services are inadequate and inconvenient. Furthermore, on a trip-by-trip basis cars appear less expensive than public transport – especially when household members travel together – because road users do not pay directly for some of the biggest costs (for example, the value of urban land under most roads and many public parking areas) and because major car-related expenses such as registration and insurance have to be met regardless of car use and they have to be paid only once per year.

Higher urban density, which makes trips shorter and makes public transport infrastructure more economically viable, is therefore one of the keys to greater use of public transport. Figure 7.1 draws on data collected by Peter Newman and Jeffrey Kenworthy from scores of cities in four continents to relate the overall population density of these cities to their petrol consumption, which is representative of their reliance on private motor vehicles and their carbon dioxide emissions from transport. They obtain similar results from comparing local government areas within the same city.[14] The less dense the city or local government area, the higher its per capita petrol consumption. Denser cities and denser local government areas can and do rely to a far greater extent on trains, trams, ferries, cycling and walking. This is the case in most European and a number of Asian cities, and indeed for local government areas within US and Australian cities. Based on these extensive data sets, Newman and Kenworthy find that the transition to much greater automobile dependence occurs consistently when the population density decreases below 30 to 40 residents and jobs per hectare.[15]

Newman and Kenworthy explain the relationship in Figure 7.1 in terms of the average time that people are prepared to spend on journeys within a city. They identify two characteristic timescales: 10 minutes of walking, corresponding to about 1 km of distance, that is the accepted time that people take to go to public transport or a local amenity; 30 minutes that is the average travel time budget for people walking and/or taking public transport to urban services and particularly their jobs in an urban subcentre. These travel time characteristics are then taken as the basis for improving urban design to reduce automobile dependence, by restructuring a city into clusters of residents and jobs.

## Clusters

Newman and Kenworthy restate the problem thus:

> A large disparity between where people live and where they
> work means that trip distances are longer in US and Australian

cities than in European cities. However, the high density of jobs located in central areas of the former is also the backbone of the public transportation system in these cities. Thus, simply moving jobs from central areas to the suburbs is not going to guarantee any less travel overall – unless they are located proximate to high density areas and thus maintain transit viability.[16]

A major part of the solution is to restructure an automobile city into a hierarchy of population clusters. At the top level of the hierarchy the city is composed of a series of 'transit cities' of diameter 20–30 km. The town centre of each transit city would have a radius of about 3 km, the distance that could be covered in about 30 minutes by a brisk walker. Thus it would be of similar dimensions to the walking cities that existed in Europe and Asia before the industrial age and would contain 100 000 residents and jobs. Town centres would be linked together and to the central business district by fast public transport, preferably heavy rail.

Each transit city would contain, in addition to the town centre, a number of local centres, each with a radius of about 1 km, corresponding to a 10-minute walk, and containing up to 10 000 residents and jobs. Within a transit city people would travel between local centres, and between local centre and town centre, by bus, light rail or cycling, depending upon distance and physical ability.

Figure 7.2 shows a conceptual plan by Newman and Kenworthy for reconstructing a hypothetical city according to these principles.[17] The distribution of transit cities appears to correspond to the existing subcentres of Sydney, Australia. An early version of the idea of a hierarchy of clusters, without the characteristic distances, was presented in the book, *Seeds for Change* by Deborah White and colleagues.[18]

There is already a body of theory and practical experience with *local centres*, sometimes called *compact developments* or *urban villages*. They are discussed in more detail by the Victoria Transport Policy Institute[19] and in many of the writings of Newman and

FIGURE 7.2 **A conceptual plan for reconstructing an automobile city**

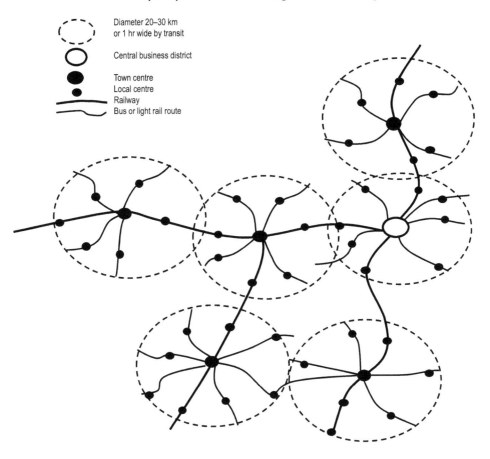

Diameter 20–30 km
or 1 hr wide by transit

Central business district

Town centre
Local centre
Railway
Bus or light rail route

SOURCE Newman and Kenworthy (2006).

Kenworthy.[20] Such local centres are popular in European cities and are starting to appear in North America and even China where they tend to have much higher population densities than in western countries. Key characteristics of local centres are:

- Mixed land use, with commercial offices and shops on main spines surrounded by residential uses.
- High-density compact structure, with every part of the cluster within walking distance.
- Sufficient population for community facilities, such as shops, an elementary school, professional services, and offices.

- Public spaces, with strong design features (water, street furniture, playgrounds etc)[21] and car access restricted. Parking is usually underground or around the periphery of the cluster.
- Linked to the rest of the city by excellent public transport, preferably light or heavy rail.

In addition they may also have some of the following specific features:
- extensive landscaping, including gardens on top of buildings and on balconies, and fruit and vegetable plots
- a mixture of public and private housing, with an emphasis on family dwellings and thus quite large internal spaces
- extensive provision for children, within easy view of dwellings
- community facilities such as libraries, child care, aged centres and, in a few cases, small urban farms
- pedestrian links with car parks placed underground and traffic calming on any peripheral roads.

Many European cities are fortunate in their evolution, starting as compact walking cities, then expanding the inner city zone by means of light rail and metro rail, expanding outwards along radial pathways with heavy rail and finally adding car-dependent outer suburbs. Thus these cities already have a basic structure that fosters walking, cycling and public transport and hence requires low per capita car use. Nowadays many European cities are models for the more difficult transformations needed in North American and Australian cities and those Asian cities that are car dependent. For instance, the new solar-efficient suburb of Vauban in Freiburg (Germany) can be accessed by light rail, makes car parking expensive and limits it to the suburb's perimeter. New housing developments in Stockholm (Sweden) must be located within walking distance of a train station, and major expansions are planned for urban-suburban rail and county buses.[22] Many positive examples of resilient cities are given in the writings of Peter Newman and colleagues cited in this chapter.

## Breaking down car culture

The traditional approach to transport planning – building or widening major roads and providing large numbers of parking spaces in urban centres and subcentres – was done under the erroneous belief that this is an effective means of alleviating traffic congestion. In practice, after a brief initial reduction in congestion, the new and improved roads simply attract more vehicles onto the roads. Similarly, providing more parking spaces at major trip destinations simply encourages more people to drive. This phenomenon of induced traffic growth is now recognised by many urban planners. They have responded by closing traffic lanes on major roads and replacing them with light rail or cycle lanes or pedestrian zones; placing congestion charges on entry to city centres (eg, in London, Stockholm and Singapore); reducing the numbers of parking spaces in centres and subcentres while increasing parking spaces for cars and bicycles at suburban train stations; and charging prices for the remaining parking places that reflect land values.

Other strategies available to reduce the incentives to drive include:
- providing shuttle buses to and from major public transport nodes
- printing transport maps on tickets and advertising material
- offering rewards for public transport users
- providing networks of safe bicycle lanes
- ensuring that venues for major sporting and cultural events are well served by public transport.


## Social psychology of travel behaviour

A vital element that is currently reinforcing car culture is the social influences, the transport-related messages that people receive from advertising, the media, films and peers. These messages create the impression that the car is the normal and most desirable means of transportation. To encourage public and decision-maker support for a shift away from cars to public and active transport, the dominance of these messages must be eroded. Drawing upon sociological and psychological literature related to social influences on behaviour, PhD candidate Jacqueline Hicks has analysed these messages, developed a conceptual framework and applied it to the problem of encouraging people to choose public and/or active modes of urban transport instead of cars. She recommends future strategies that take into account the way messages portray images of transport use; the prevalence of transport use; the proximity of people to the use of different modes of transport; the framing of transport problems, solutions and responsibility; the portrayal of a sense of fairness and power; and the way messages lead to competency to deal with the complexity and variability of transport. She also applies to travel behaviour some of the insights gained from the successful public health campaign to reduce smoking in Australia.[23]

## Improved vehicles and fuels

### Improved vehicle technologies

Over the past decade or two significant improvements have been made to the design of motor vehicles to reduce their fuel consumption per kilometre through more streamlined shapes and fuel-efficient engines. Mandatory fuel economy standards are needed to push this process further.

However, the potential for further efficiency improvements is limited and anyway all the cumulative improvements to date have been swamped by the increased numbers of vehicles on the road. While vehicle use can be reduced by the policies suggested in 'Integrated urban and transport planning' above, there will always be a need for some motor vehicles, even in cities with excellent facilities for public and active transport. For some trips, some purposes and some people there are no practical alternatives to the motor vehicle for personal transport and freight deliveries.

At present the most promising urban-suburban solution in this situation comprises hybrid, plug-in hybrid and all-electric vehicles charged with renewable electricity. Hybrid vehicles have two or more separate methods of propulsion, usually an internal combustion engine running on liquid or gaseous fuels and electric motors powered by batteries. The electric motors are used mostly during stop-start driving in cities, where they are much more efficient than petrol or diesel engines. The batteries of hybrid electric vehicles (HEVs) are charged by the petrol or diesel engine when the vehicle is cruising at a steady speed and by regenerative braking, which captures some of the energy that would be lost when the vehicle decelerates. The plug-in hybrid electric vehicle (PHEV) has greater range than the simple HEV, obtained at a price by larger battery storage. These batteries can be recharged by plugging them into grid electricity when the vehicle is parked as well as the methods of charging HEVs. The Chevy Volt is a PHEV with a long range; its petrol engine takes over seamlessly when the range of the batteries is reached. For the majority of urban/suburban drivers who rarely exceed 50 km per trip all-electric vehicles (EVs) are becoming popular. Indeed, in the USA in the first half of 2013 their sales exceeded those of PHEVs.[24] Despite their limited range, EVs have the advantage of the mechanical simplicity of an all-electric propulsion system, are often an appropriate choice for a second family car and are also being purchased as the single car by families and individuals who choose to make their rare long distance trips by hire car or public transport.

In countries with very high contributions from coal-fired power stations in the grid electricity mix, charging electric vehicles from the grid may give little or no reduction in GHG emissions compared with using a petrol or diesel car. This is currently the case in Poland, South Africa, mainland Australia and China. One solution is to purchase Green Power, where such schemes exist. This involves paying a little extra to electricity retailers for certified renewable electricity.

Policy support needed for EVs and PHEVs from local

government and business is the provision of charging points that sell renewable electricity in car parks in town centres and subcentres. This is a classic chicken-and-egg problem: in the absence of a network of charging points, most people will not purchase electric vehicles, but until there are sufficient electric vehicles on the road, governments and businesses will tend to provide the bare minimum of charging points.

For long journeys along roads where there are no fast charge or battery exchange facilities for EVs, biofuels seem to be the only alternative at present. Government incentives are needed for research, development and demonstration projects for efficient, ecologically sustainable biofuels production.

### Liquid and gaseous fuels from renewable sources

For fuel-based motor vehicles, stationary engines and gas turbines, energy is stored in the chemical bonds of the fuel. The fuels can be fossil – eg, petrol, diesel, liquefied petroleum gas (LPG), or liquefied natural gas (LNG) – or renewable fuels from biomass – such as ethanol, methanol, butanol, biodiesel or biogas. An important characteristic of a fuel or other energy storage medium is its energy density, measured either as the energy stored per kilogram (kg) or per cubic metre ($m^3$) of the store. Keeping the latter as high as possible is important for vehicles, less so for stationary energy. As shown in Table 7.2, ethanol has significantly less energy density than petrol. Methanol has only about half the energy density of petrol, but this disadvantage is partially offset by the fact that methanol can be burned more efficiently in engines. Indeed it fuels some racing cars.

Liquid and gaseous biofuels can be readily stored for months or even years without significant standing losses. In these forms, biomass can be used for electricity generation, industrial heat, stationary engines or motor vehicles. The principal constraint at present is that there are limited biomass residues from agriculture and forestry, considering the big demands likely to be made on

TABLE 7.2 **Energy densities of different liquid and gaseous fuels**

| Fossil or renewable | Fuel | Energy density | |
|---|---|---|---|
| | | MJ/kg | MJ/L |
| Fossil | Petrol, diesel, kerosene | 46 | 33 |
| | Liquefied petroleum gas | 50 | 25–28 |
| | Natural gas at atmospheric pressure | 53 | 0.038–0.400 |
| | Compressed natural gas | 53 | 8–11 |
| Renewable | Methanol | 20 | 16 |
| | Ethanol | 30 | 23 |
| | Biodiesel | 40 | 34 |
| | $H_2$, gas | 120 | 1.9–2.7 |
| | $H_2$, liquid | 120 | 8.7 |
| | $H_2$, metal hydride | 2–9 | 5–15 |

SOURCES BREE (2012) and Sørensen (2011), p 565.[25]

these residues for road and air transport, manufacturing chemicals including fertiliser, and steel-making. The present first generation method of producing biofuels such as ethanol from the fermentation and distillation of sugars is inefficient and so, if implemented on a large scale, would compete with land for food production. Second generation production of biofuels from the ligno-cellulosic components of plants is much more efficient in terms of yield per hectare, but is not yet a mature industry.

Hydrogen and methanol are energy carriers, not primary energy forms, so the method of production is critical for a low-carbon energy system. At present the principal means of producing hydrogen from RE is via the electrolysis of water, a process with efficiency of about 70 per cent. Research is in progress on alternative low-carbon, low-energy methods of producing it, such as from algae and genetically engineered bacteria. Since hydrogen has low volume density at atmospheric pressure, it has to be compressed in order to be stored in a fuel tank of moderate size. Commercial gaseous hydrogen containers at present can hold pressures of 20–30 million Pascals (MPa), corresponding to energy densities of roughly 2–3 megajoules per litre (MJ/L), which is only a few per cent of oil's energy density. Recent research suggests that it may be possible to increase the pressure to 70 MPa by using high-strength composite materials such as Kevlar fibres. Hydrogen stored in liquid form in a cryogenic tank has a higher volumetric energy density, but it is still less than petrol's and, in addition, requires a much heavier fuel tank. Because of the energy required to compress or liquefy the hydrogen gas, the EE of the whole process of producing and storing hydrogen is much lower than that of petrol. All the other forms of hydrogen storage for transportation have much lower energy densities per cubic metre and so are unsuitable for small vehicles (that is, cars and motor bikes). They may, however, be used in large vehicles, such as trucks and buses, and stationary sources of energy.[26]

The most efficient way of using hydrogen as a fuel in large vehicles is to combine it with oxygen from the atmosphere in a chemical reactor called a fuel cell to produce electricity (plus water and heat) and to propel the vehicle with electric motors. Fuel cells are 40–60 per cent efficient, compared with the internal combustion engine of about 25 per cent. However, they are still at the early diffusion stage of technological maturity and are expensive. Limited numbers of fuel cell buses are deployed in several countries, but no fuel cell cars. Clearly the large-scale use of hydrogen as an RE carrier is not close.

Methanol has some advantages over hydrogen as a motor vehicle fuel. Indeed methanol produced from the methane in natural gas is conventionally burned as a high-octane fuel in some racing cars. It can also be used in ordinary cars with slight modifications to the engines and fuel lines. It is more toxic, acidic and corrosive than ethanol, it requires larger fuel tanks per unit of energy, but is still amenable to 100 per cent replacement of petrol. However, producing methanol from fossil fuels is unacceptable in terms of GHG emissions. Fortunately the oldest method of production is from an RE source, wood – indeed bio-methanol used to be called 'wood alcohol'. It can be produced in a low-carbon process by gasifying biomass to produce a synthesis gas, a mixture of carbon monoxide (CO) and hydrogen ($H_2$). These are reacted together over a catalyst to form methanol. With methanol in your fuel tank you can in theory convert it directly into electricity in a methanol fuel cell. However, this is inefficient and, like the hydrogen fuel cell, expensive. The waste products emitted by the methanol fuel cell are $CO_2$ and water. However, the $CO_2$ absorbed during photosynthesis when the biomass regrows offsets this, provided the rate of use of methanol is not greater than the rate of sequestration of the $CO_2$.

Australian scientists Barney Foran and Chris Mardon have developed a scenario in which forests are grown on marginal land for the large-scale production of bio-methanol and/or bio-ethanol for motor vehicles. The scenario combines land remediation with fuel production and deserves serious consideration. Deep-rooted perennial crops would be grown to mimic the function of ecosystems that are now gone or non-functional. They would halt the leakage of water and nutrients that nowadays contributes to dryland salinity and soil acidification in Australia. As well as liquid fuels, valuable co-products could include chemicals, plastics, lubricants, solvents, paints, building materials, and starch and protein by-products.[27]

## Intercity transport

For intercity trips, high-speed rail (HSR) has the advantages that it links city centres, gives more space to passengers, allows frequent use of mobile phones (except in carriages designated as 'quiet'), gives access to WiFi and exposes travellers to much less ionising radiation from cosmic rays than air travel. For distances up to about 600 km, door-to-door travel times are often comparable. HSR is widespread in Japan, China and Europe, but investment in HSR is long overdue in North America and Australia.

## Encouraging signs of change

As is the case with energy demand and supply, there are early signs of change towards a more sustainable development pathway. Although the use of cars continues to grow on a global scale, the rate of growth has been declining steadily, decade by decade, in developed countries. Drawing on data from their Global Cities Database, Newman and Kenworthy have shown that the rate of growth of vehicle-km per capita in developed country cities has decreased from 42 per cent in the decade 1960–70 to 5 per cent in the decade 1995–2005.[28] In US cities, which have the world's highest per capita car use (see Figure 7.1), per capita car use began to plateau in 2000 and declined in absolute terms since 2005,[29] while public transport use has increased. In Australian cities, the second highest per capita car users in the world, the decline commenced in 2003–04. In both the USA and Australia the decline predated the rise in petrol prices in 2007 and 2008. The provision of new and improved rail services is likely to be a factor in Australia.[30]

In some of the major cities of rapidly developing countries, unsustainable and sustainable development are now proceeding side by side. While car use is increasing rapidly in these cities, the congestion, noise and air pollution are becoming intolerable to the citizens. This is motivating rapid growth in public transport. In China 82 cities are building metro railways and dozens of cities have added light rail. The Shanghai metro has grown to 13 lines, mostly built since 2000 and now carrying 8 million trips per day. High-speed intercity rail is expanding rapidly in China and, as a consequence, some air links between these cities are closing down. This is encouraging because, in terms of GHG emissions, domestic air travel may have to be dispensed with in all but the largest countries. India is building metros in 16 cities. Middle Eastern cities are building rail for the first time.[31]

Nowadays a growing number of urban planners recognise the need to increase population density, day and night, around public transport stations. Indeed, transit-oriented development simultaneously increases the number of public transport passengers while providing a source of finance for public transport. However, there is resistance to increased densification from homeowners who have become accustomed to the luxury of owning large blocks of land in inner city locations.

## Conclusion

Many of our cities are big contributors to unsustainable development. They drive (literally) climate change; pollution of air, water, soil and associated illness; loss of valuable land to roads and parking; and social isolation. They impose high economic costs in providing infrastructure at their far-flung fringes. They are vulnerable to the current peak in global oil production. The prime cause of these problems is dependence upon the motor car. However, with community support and political will, they can be transformed into healthy, liveable environments.

We can and must build on recently sprouting seeds for change to renovate our cities so that people take fewer trips and the majority of trips are by public transport, cycling and walking. To do this we must reject the notion promulgated by some neoclassical economists that urban planning (and most other decisions) should be left to the market. Cities must be planned, with substantial input from citizens. Urban planning and transport planning must be fully integrated, with increased population density at public transport nodes and stations, and fewer freeways and inner city parking places. A feasible pathway to urban reconstruction is to plan and implement the transformation of cities into a hierarchy of clusters, as outlined in 'Integrated urban and transport planning' above and Figure 7.2. Public and private urban transport should be fuelled by renewable sources of electricity.

A key early step in the transformation must be to address the heavy conditioning, to which we are all exposed, that cars are the normal means of transportation in modern societies.

For rural transportation and international air travel, second generation biofuels offer promise, subject to the ecological sustainability of their life-cycles and further development of the fuel production processes.

# PART C:
# POLICIES, STRATEGIES, POLITICS AND ACTIONS

# 8: GENERAL POLICIES FOR THE GREAT TRANSITION

> We are now in the midst of a fight between the past and the future.
>
> Christine Milne[1]

Which country, you may ask, is most advanced in making the Great Transition to a sustainable energy system? Which policies have enabled it to reach its present position and are driving it further along the sustainable development pathway? It all depends on your perspective.

- Iceland already has 100 per cent renewable electricity and much renewable heat. Its ample natural hydro and geothermal resources have enabled the market to achieve this system. However, Iceland needs policies to transform its road transport system, which is still based on fossil fuels.
- Although Denmark has no hydro or geothermal resources, it already generates 30 per cent of its electricity from the wind, a significant fraction of its electricity and residential heat from agricultural residues and it has an excellent public transport and active transport system. It has targets to achieve 100 per cent renewable electricity and heat by 2035 and carbon-free transport by 2050 backed up with regulations and standards for energy efficiency in buildings, feed-in tariffs for renewable electricity, and membership of the European Union (EU) emissions trading scheme.

- Following a policy of energy diversity China is building coal-fired power stations like there is no tomorrow. At the same time it has the largest installed capacity (in gigawatts) of wind power in the world, is the biggest producer and user of solar hot water systems and the leading manufacturer of solar photovoltaic (PV) modules. Its specific policies include strong targets for renewable energy (RE), feed-in tariffs for wind energy, protection and promotion of RE industries in their early stages of development, mandatory standards for energy consumption per unit of product in key industries, pilot emissions trading schemes in several cities, rapid development of urban rail and inter-city high-speed rail, and funding of research and development.

The practices and experiences of these and other countries in fostering sustainable energy suggest the need for a wide range of policies. This chapter addresses general policies that are not technology specific and Chapter 9 addresses policies targeting specific RE technologies. However, first we must face the magnitude of the challenge.

## Barriers to change

Huge multinational corporations (the self-styled Greenhouse Mafia) and their supporters in government, the public service, professional organisations and the media stand in the way of the Great Transition. They see the efficient use of energy as a threat to their vested interests, rather than an opportunity for the advancement of society and the environment. When they claim falsely that the only choice for electricity supply is between fossil fuels and nuclear power, they are really saying that we can choose between an industry dominated by BHP Billiton and Rio Tinto on one hand and by BHP Billiton and Rio Tinto on the other. The fossil fuel and nuclear industries exercise enormous political power, both directly through lobbying government and indirectly through the media.

They receive huge subsidies to increase the profits of their well-established technologies (see Ch 6, 'Economics' and Ch 9, 'Removing subsidies to fossil fuels and nuclear energy'). They benefit from an institutional structure that supports the sale of energy rather than energy services (see Ch 4, 'Definitions' and Ch 4, 'Overcoming barriers: strategies and policies') and an infrastructure that favours large centralised power stations and an extensive system of roads and road funding. They ensure that education, training, information and research are directed predominantly towards their own industries, while disseminating misinformation about renewable energy (RE) and lobbying to delay or water down energy efficiency (EE) standards.

The barriers are not simply those of bias to cheap and nasty technologies. As demonstrated by the identity $I = PAT$ introduced in the Introduction, growth in population and consumption per person must also be halted. Underlying these drivers of unsustainable development is a flawed economic system that fosters endless economic growth on a finite planet and the associated culture of consumerism. This concern is, or was, shared by Nicholas Stern, former chief economist of the World Bank and author of the *Stern Review on the Economics of Climate Change*, who has been quoted as saying that, in order to combat climate change, the rich countries will need to abandon economic growth.[2] However, he has not repeated that statement publicly and more recently has returned to the promotion of economic growth.

A policy is a statement of intention. To be effective, it must be coupled with a strategy and process for implementation and enforcement. Anthropogenic (human-induced) climate change is an outcome of the interaction between two complex systems, Earth's climate and the human socio-economy. It is true to say that 'to every complex problem there is a simple solution ... and it is wrong'. No single policy measure is sufficient to mitigate climate change or even to implement the partial solution of driving the Great Transition to a sustainable energy system. This is partly due to the diversity of barriers to change, partly to the fact that energy policy affects many sectors of the economy and partly to the great diversity of sustainable energy technologies and measures at different stages of maturity and various scales of manufacture and use. Different technologies on different scales at different stages of maturity need a wide range of policies, as discussed in Chapter 2, 'Maturity of energy technologies' and Chapter 9. Different scales of technology – small-scale for residential use; medium-scale for commercial, local community and small industrial use; and large-scale for large industrial use and public infrastructure – need targeted policies.

This diversity of barriers, economic sectors, technologies and scales entails that several types of policy must be designed and implemented, namely:

- economic instruments, such as a carbon tax, emissions trading, feed-in tariffs, tradable certificate schemes, low-interest loans, grants, fees, rebates and subsidies
- laws, regulations, standards and labelling
- research, education, training and information
- institutional change, for instance for energy service companies to replace energy supply companies, as discussed in Chapter 4
- planning and design for cities, buildings and other products
- support for community participation and the development of community sustainable energy projects
- population stabilisation
- industry policy to foster the growth of strategic industries for a low-carbon future
- restructuring the economic system into one that does not foster growth in the use of energy, materials and land.

Within each policy type a range of policy options exist. For convenience, general policies for climate mitigation, that is, those that are not technology specific, are discussed in this chapter and policy instruments designed for specific technologies of various scales are discussed in Chapter 9. In addressing the general policies we first consider the strengths and limitations of economic instruments involving a carbon price and then policies to address the two non-technological drivers of climate change and environmental impact in general: population growth and economic growth, or more precisely, growth in the use of materials, energy and land.

## What is a carbon price?

Neoclassical economists naturally favour economic instruments over non-economic instruments. Furthermore, within the broad category of economic instruments, they generally prefer market instruments, such as a carbon price, as opposed to non-market instruments, such as grants, fees, rebates and subsidies. This is not surprising, because one of the key value judgements underlying neoclassical economics (described in Box 8.1) is that decisions are best made in markets.[3] The justification of this stance is that, in theory, market instruments are more economically efficient (that is, less expensive for the same outcome) than non-market economic instruments and regulatory measures for reducing emissions. The standard explanations of this advantage are that:

- markets give firms and individuals more flexibility in deciding how and when to make emissions reductions and
- markets give priority to the lowest cost actions.

However, we shall see that in practice market instruments, designed by governments under pressure from vested interests, can be expensive for consumers and almost ineffective in environmental terms, as demonstrated by the first two stages of the European Union (EU) emissions trading scheme (ETS) discussed below. With this preamble, let's consider the category of market instruments most favoured by neoclassical economists.

A carbon price is a generic term for an economic value placed on the emission of greenhouse gases (GHGs) into the atmosphere due to human activity. A carbon price usually takes the form of either a carbon tax or as the price of permits in a cap-and-trade ETS. The price is designed to create an incentive to reduce

**Neoclassical economics**

Neoclassical economics focuses on how idealised markets determine prices, outputs and incomes. It seeks to show how the interaction of supply and demand creates market-clearing prices. Demand comes from individual consumers who are assumed to be seeking to maximise their utility (or satisfaction) through acquiring goods and services. The supply of those goods and services comes from firms who are assumed to be seeking to maximise profits. The neoclassical view represents markets as efficient and want-satisfying, just so long as competitive conditions prevail and prices are allowed to continually adjust to small (or marginal) changes in demand and supply. There is ongoing controversy about whether this neoclassical theory adequately represents how markets actually work in the real world.

Neoclassical theorists sometimes acknowledge that market prices do not properly reflect all the relevant social costs. 'Externalities' are said to exist, for example, where environmental damage results from self-interested market behaviours. Some neoclassical economists concede that, in these circumstances, there is a need for governments to 'intervene' in markets to adjust prices, eg, by imposing a carbon tax to create a financial disincentive for environmentally degrading activities. Critics of neoclassical theory tend to be more sceptical about the adequacy of this incremental pricing approach in dealing with great challenges like the prevention of climate change. General critiques of neoclassical economics have been made on the grounds of inadequate treatment of environmental and social justice issues, internal inconsistency, unrealistic value-laden assumptions (despite claims of 'objectivity') and its use as the basis of a political ideology.[4]

emissions. It can also be considered as a means of including all or part of the costs of the environmental, health and economic damage caused by burning fossil fuels in their prices.

Ideally the carbon price, expressed in terms of a cost per tonne of emissions, should increase over time, gradually shifting energy generation, industries and businesses towards low emissions. Scenario studies suggest that, to achieve zero net emissions in some countries, it may be necessary to increase the carbon price to $100/tonne or more over a period of a few decades. This appears to be a formidable impost until one realises that, by the time the price has reached such levels, EE programs are likely to have cut in half the energy needed to perform most energy services and most of the remaining energy use will be supplied by renewable sources. The price per tonne of GHG will be high, but the number of tonnes emitted per energy service will be much less than at present and so costs will be contained.

A *carbon tax* sets a definite price on emissions and allows the market to adjust the quantity of emissions. Thus it determines the economic cost of reductions at the risk of potentially high emissions if the market and complementary measures are slow in developing emission reduction mechanisms. On the other hand, a *cap-and-trade ETS* fixes the quantity of emissions (the cap) and allows the market to adjust the price of emissions (the trade). Thus it determines the quantity of emissions at the risk of high costs if the market and complementary measures are slow in developing emission reduction mechanisms.

Before comparing these two principal market mechanisms for cutting GHG emissions, we consider whether a carbon price is necessary for the mitigation of human-induced climate change and, if so, whether it is sufficient.

## Is a carbon price necessary? Sufficient?

A low carbon price, whether it is in the form of a carbon tax or an ETS, may not be high enough to change people's energy-using behaviour. In the jargon of economics, the price elasticity of demand is too low. However, even a low carbon price sends a message to investors that they would be taking a financial risk to invest directly or indirectly in fossil fuels, whether the proposed investment is a new coal-fired power station, oil refinery or motorway. Even a low carbon price has institutional support and so can be easily increased in the future, thus increasing the financial risk to investors in fossil fuels and related products and services. In my view this message is the key role of a carbon price. Although the carbon risk message has not yet been taken seriously by many investors, air pollution is having an impact. In the USA a survey by Associated Press reported that more than 32 mostly coal-fired power plants will close and another 36 plants could also be forced to shut down in 2014–15 as a result of new Environmental Protection Agency (EPA) rules regulating air pollution.[5] In order to increase financial risk alone, I believe that a carbon price is necessary. However, there are additional reasons.

A carbon price applied 'upstream' – for example at coal mines and oil and gas wells – flows 'downstream' to all economic transactions regarding products and services based indirectly on fossil fuel use. A big energy user, such as an aluminium smelter using predominantly electricity from fossil fuels, would have to pay higher prices for electricity, as would a purchaser of a drink in an aluminium can. Thus it is far more efficient, in terms of reduced transaction costs, to apply a carbon price upstream than to attempt to calculate carbon prices for millions of individual downstream products or to regulate their sale individually.

Furthermore, consumers of greenhouse-intensive products and services that are either high energy-users or have high quantities of embodied energy receive inconsistent messages, if on one hand there is an education campaign and strict regulations limiting the use of these products and services, while on the other hand those products are very cheap. Different policies must work together, not in opposition.

Another justification for a carbon price is to raise revenue for the government to help fund a just transition to a sustainable energy future. A national government can potentially raise billions of dollars per year to assist households and employees disadvantaged by the transition and to pay for new infrastructure such as railways and transmission lines.

Finally it should be recognised that the carbon price is not really an additional impost, because people are already paying indirectly for the many impacts of anthropogenic climate change. Heatwaves, droughts, wild fires, floods, more frequent severe storms and sea-level rises impose costs, which are paid for through taxes and through the suffering of people who are not covered by remedial government expenditure. A carbon price merely internalises some of these costs in the prices of the products causing the impacts. In the language of economists, it is a Pigovian tax. In the language of environmental managers, it is an application of the Principle of Polluter Pays. In practice the polluters usually pass on the costs to consumers of their polluting products and services. So, if it is high enough, the carbon price nudges high emitters and consumers of their products and services towards less polluting production processes, products and services.

Some neoclassical economists claim that a carbon price is not only necessary but is also sufficient. In other words, once a carbon price has been implemented, all other carbon mitigation policies become redundant. Indeed, many of these economists go even further by claiming that 'complementary' policies, such as feed-in tariffs and RE certificate schemes, actually make the response less 'efficient' economically speaking, that is, more expensive for achieving the same environmental outcomes. Their argument is that there is competition between the carbon price and RE promotion. Because the carbon price is supposed to be more economically efficient than specific policies to promote RE, this competition pushes up the cost of climate mitigation. In the case of an ETS (see 'Cap-and-trade emissions trading', below), the expansion of RE reduces the consumption of fossil fuels, thus reducing the demand for emission permits and so also reducing the price of these permits. This in turn reduces the incentive to lower emissions.

Although this specific problem is real, it can be solved by progressively tightening the cap on emissions, that is, reducing the total number of emission permits on the market. This requires better coordination between emissions trading and policies to promote RE.[6]

The fundamental flaw in the general argument that an ETS is the most efficient instrument is that it is based on the unrealistic assumption that energy supply and demand occur in a perfect competitive market. There are several reasons why this assumption is invalid:

- Producers and consumers have incomplete information. This is especially true of residential electricity consumers, who are often faced with price structures that do not reflect the actual costs of supply by time of day and by the maximum demand they receive.
- The market rarely reveals the huge subsidies granted to fossil fuel production and use and to nuclear power. These subsidies dwarf the small, declining subsidies to RE (see Ch 6, 'Economics' and Ch 9, 'Removing subsidies to fossil fuels and nuclear energy').
- A carbon price, especially one determined by emissions trading, does not take account of all the benefits of RE additional to climate mitigation, such as increased energy security (including increased security against increases in fuel prices), reduced pollution of air, water and soils, and reduced water use.
- The market rarely provides expensive infrastructure such as transmission lines, railways and networks of bicycle paths and pedestrian paths. This usually depends on political decisions by governments of various levels. A carbon price may help sway the decision, but it rarely overrides budgetary factors and political influences.
- The market fails badly for both EE and RE in homes and offices where there are generally split incentives, such as between landlords and tenants. Without government intervention, for example through regulations and standards, it is neither in the interest of the landlord nor the tenant to make a building envelope energy efficient (see Ch 4, 'Barriers to energy efficiency').

- In the absence of government incentives, the market rarely funds research, development and demonstration of new technologies.
- The market in general, and cost–benefit analysis in particular, do not handle well the different physical risks of various technologies. For instance, it might value equally a new nuclear power station and a group of concentrated solar thermal power stations with thermal storage, having the same energy output and reliability, despite the fact that the nuclear station imposes a small probability of a devastating accident and is never insured sufficiently to cover such events (see Ch 6).
- The market, based on marginal economics, takes a short-term view of technological change. So, a low carbon price might favour the replacement of a proposed coal-fired power station with gas-fired, which still has an unacceptably high rate of GHG emissions, instead of supporting the long-term solution that requires a direct transition to RE. Although most RE technologies are more expensive at present, they are very likely to become cheaper over the next few decades, while fossil fuels are likely to become dearer. Thus a decision at the margin is not generally the best decision, even in economic terms, when a radical long-term system change is required.
- Policy failure in the implementation of a carbon pricing system, especially an ETS, can turn it into a very different system from the idealised one considered by neoclassical economists to be highly efficient in economic terms. For example, the European and Australian ETS schemes have exemptions for major GHG polluters, allow polluters to purchase cheap overseas offsets of dubious effectiveness, have modest targets determined by political forces and negotiations rather than economic efficiency, and are unlikely to be able to internalise environmental and non-environmental externalities adequately.[7]
- In particular, if the carbon price is very low as the result of policy failure, it will be incapable of making any change in the energy system, let alone a radical change. This was the

situation in Europe in April 2013, when the carbon price fell to €2.8 per tonne of $CO_2$ after the EU parliament rejected an emergency proposal to reduce the supply of carbon permits and hence push up their price. More generally, if policy failure leads to erratic variations in the carbon price, investor confidence is undermined and complementary measures are needed to restore it.

Hence, the notion that a carbon price alone is sufficient to drive the transition to a sustainable energy system is poorly based. In the words of Nobel Prize-winning economist Joseph Stiglitz, 'markets by themselves do not produce efficient outcomes when information is imperfect and markets are incomplete ... or when technology is changing ...'[8] Alternative technologies, industries, jobs, infrastructure and lifestyles must be put in place, actions that require a wide range of policy instruments additional to pricing. Otherwise consumers will end up paying the carbon price without being able to make any change in their energy-use behaviour. It's pointless asking householders to reduce their use of coal-fired electricity to heat their homes if they live in rented accommodation with no insulation, no solar access and no connection to natural gas. It's pointless asking commuters who live on the outskirts of cities to reduce their car use if they are poorly served by public transport. Therefore we need a portfolio of mutually reinforcing policies, including a carbon price as one element, to give people options to change behaviour. Next we consider the principal types of carbon pricing.

## Carbon tax

'Carbon tax' in general is a shorthand terminology for a tax on GHG emissions, including emissions from GHG that do not contain carbon, such as nitrous oxide ($N_2O$).

A carbon tax is an indirect tax – a tax that is levied on 'goods' and services rather than on people, on expenditures rather than on incomes. However, unlike the usual kind of Goods and Services Tax, a carbon tax has a differential impact on the prices of goods and services according to the amounts of fossil fuels used in their production.[9] A tax on $CO_2$ emissions puts the biggest price increase on the most greenhouse-polluting fossil fuel, coal, and products made from it, and the smallest price increase on the cleanest (in greenhouse terms) fossil fuel, natural gas, and the products made from it. (However, recent studies find that life-cycle emissions from shale gas are comparable with those from coal when evaluated over a 20-year time horizon.)[10] Specifically, a $CO_2$ price that adds $1 per gigajoule (GJ) to the price of coal would add $0.84/GJ to oil and $0.57 to natural gas. However, calculating the carbon price per unit of energy content is only the first step – the efficiency of conversion must also be taken into account. For example, for generating electricity, natural gas can be burned with a conversion efficiency of about 60 per cent in a combined-cycle power station, while the best efficiency of a modern coal-fired power station is about 40 per cent, so in this case the relative $CO_2$ impost between electricity from coal and from natural gas becomes $60/(57 \times 0.4) = 2.6$. So a carbon price that adds $1/GJ to the price of coal would add only $1/2.6 = $0.38 to the price of natural gas in a combined cycle power station.

As discussed in Chapter 5 'Energy payback', most RE technologies have low fossil fuel inputs to their manufacture and installation, compared with their lifetime energy generation, and emit no GHG emissions from their operation. Hence they have very low life-cycle $CO_2$ emissions, and so negligible carbon taxes would be embodied in the prices of energy they generate. Two RE technologies are exceptions to this general observation. Hydroelectric schemes that flood extensive vegetated valleys emit methane, a potent GHG, from anaerobic decomposition of the vegetation. Some forms of bioenergy have large inputs of fossil energy and hence $CO_2$ emissions via processing, fertiliser and transportation over long distances; they may also have net emissions resulting from the clearing of dense native forests and replacing them with less dense plantations (see Ch 5).

Key design issues for a carbon tax are the GHGs and fuels covered; the tax rate and how it will evolve over time; the collection point of the tax (how far upstream); its relationship to existing taxes; any exemptions or concessions granted to certain stakeholders; and how the revenue is spent. These choices determine the effectiveness of the tax and its social impacts. Most environmentally related taxes with implications for GHG emissions in Organisation for Economic Co-operation and Development (OECD) countries are levied on energy products and motor vehicles, rather than on $CO_2$ emissions directly. Nevertheless, several European countries have an explicit carbon tax, albeit with exemptions for favoured industries. Several members of the EU even have a carbon tax in addition to participating in the EU ETS. One of the few non-European countries with a carbon tax is Costa Rica, which applies part of the tax revenue to fund incentives to property owners to practise forest conservation.

## Cap-and-trade emissions trading

Cap-and-trade emissions trading is much more complicated than a tax. It involves setting a target of permitted emissions; allocating permits or allowances to producers (either direct or indirect) of the emissions, while ensuring that the total number of permits is consistent with the target; then mandating that producers of emissions acquire sufficient permits to cover their emissions; and monitoring the process. Participants in the market may trade the permits and this sets the carbon price. Then, in theory, participants who can reduce their emissions at low cost will sell some of their permits to participants who can only reduce their emissions at high cost. Thus, the lowest-cost measures for reducing emissions will tend to be implemented first, buying time for those with high-cost measures to change their practices. This is the basis of the claim by neoclassical economists that an ETS is more economically efficient than a carbon tax or any other policy instrument. However, the claim rests on assumptions about the operation of the market for permits that are not in general satisfied in practice.

Hence the claim of economic efficiency cannot be substantiated.[11]

The basic requirements of a successful ETS are a tradable commodity, willing buyers and willing sellers.[12] To obtain willing buyers it is essential to make participation mandatory for some sections of the economy. So, despite the rhetoric of emissions trading being a market mechanism, it also requires a strong regulatory structure to be effective. Willing sellers will come forward when the target and allocation of emission permits encourages them to do so. In practice there are many ways of designing and implementing emissions trading. Important choices have to be made whether schemes are designed to:

- reduce emissions that can be physically measured, or include 'reductions' that are uncertain and estimated
- allocate emission permits free of charge to emitting industries in proportion to their current emissions ('grandfathering'), or auction permits with all comers entitled to bid
- define the liable parties (those that have to obtain the emission permits) to be the relatively small number of industries that actually produce the emissions directly, or the vast numbers of consumers who produce the emissions indirectly through their purchases of good and services
- return the revenue to people disadvantaged by carbon price
- compensate energy-intensive trade-exposed industries
- focus on $CO_2$ emissions alone, or also include several other GHGs with emissions measured in $CO_2$-equivalents
- allow banking and borrowing of emission permits
- set strong limits on the proportion of overseas offsets permitted to businesses
- set strong penalties and make-good requirements for businesses that have greater emissions than are covered by their permits.

This flexibility offers both opportunities and risks. These schemes and their rules are created by governments or international organisations, who shape them for political purposes, responding to pressures from powerful stakeholders. Furthermore, some versions of these schemes suffer inherently from market failures and they all feed into energy markets that have their own market failures. So, the design, regulation, enforcement and transparency of emission trading schemes are crucial to their success in reducing emissions.

Allocating emission permits by grandfathering gives an advantage to existing industries, many of which are big GHG polluters, and makes it more difficult for new cleaner technologies to enter the market. On the other hand, commencing emissions trading with a system that allocates all permits by auction could lead to stranded assets, big economic losses by existing providers of energy and energy-intensive products, and considerable disruption for consumers. A more politically feasible scheme is to commence by grandfathering about half the permits, but then mandate revaluations or reallocations of permits every three years (say) in which increasing fractions of permits are auctioned, until all are auctioned after (say) a decade.

Existing ETSs for climate mitigation, both operating and planned, are listed in Table 8.1. There is a wide diversity of structures and potential effectiveness. All except China have cap-and-trade schemes.

The world's first major ETS to be established was the European Union's scheme that commenced at the beginning of 2005 (see Box 8.2). It was followed in 2008–09 by the Regional Greenhouse Gas Initiative (RGGI) of 10 north-east and mid-Atlantic states of the USA.[13] The pilot ETSs commencing in several Chinese cities do not have a cap on GHG emissions but only on GHG intensity, which is emissions divided by gross domestic product (GDP). GHG intensity declines when the rate of growth of GDP exceeds the rate of growth of emissions. So if annual economic growth is 9 per cent, emissions growth of (say) 8.9 per cent would still satisfy a cap on GHG intensity. Recognising this limitation, China's powerful National Development and Reform Commission proposed that an absolute cap be placed on emissions by 2016. For the proposal to become official government policy, it still has to be accepted by the State Council.

TABLE 8.1    **Emission trading schemes by country/jurisdiction**

| Country/jurisdiction | Scheme name | Status | Comment |
|---|---|---|---|
| **EU** | EU ETS | Operating | $CO_2$; initially free permits; transport not covered. See Box 8.2 |
| **Switzerland** | Swiss ETS | Operating; planned to link to EU ETS | Similar to EU ETS |
| **Norway** | Linked to EU ETS | Operating | Similar to EU ETS, but only covers 10% of Norway's emissions |
| **Iceland** | Linked to EU ETS | Operating | – |
| **Liechtenstein** | Linked to EU ETS | Operating | – |
| **New Zealand** | New Zealand ETS | Operating | Agriculture, NZ's principal source of GHG emissions, excluded |
| **10 US states(a)** | Regional Greenhouse Gas Initiative (RGGI) | Operating | Electricity only; permits auctioned; proceeds invested in sustainable energy, etc |
| **California** | ETS; may be linked to Quebec ETS | Operating | Covers most GHG, stationary energy, transport and industry |
| **Quebec** | ETS; may be linked to Californian ETS | Operating | Similar to California ETS |
| **Japan** | ETS | Operating | Pilot in greater Tokyo region only covering $CO_2$; proposed national ETS postponed |
| **Australia** | Carbon tax → ETS | Planned to link to EU ETS | Currently has a de facto carbon tax scheduled to become ETS in 2015, but newly elected government plans to remove it |
| **Kazakhstan** | ETS | Trial | – |
| **China** | Pilot ETS in several cities and provinces | Commenced 2013 | Cap on GHG intensity, which is not a cap on emissions |
| **Republic of Korea** | ETS | Under development | From 2015; initially all permits free; excludes overseas offsets |
| **Belarus** | ETS | Under development | No details available |

SOURCE Australian Department of Industry, Innovation, Climate Change, Science, Research and Tertiary Education website,[14] with updates and comments column added by the author.
NOTE (a): Connecticut, Delaware, Maine, Maryland, Massachusetts, New Hampshire, New Jersey, New York, Rhode Island, Vermont.

## Box 8.2    The European ETS

Following ratification of the Kyoto Protocol the EU committed to reducing its GHG emissions to 8 per cent below the 1990 level by 2012. One of the policies designed to facilitate this reduction is the EU ETS, a cap-and-trade ETS that commenced at the beginning of 2005. The scheme has three phases: a 3-year pilot phase (2005–07), a second phase coinciding with the Kyoto Protocol commitment period (2008–12) and a third phase (2013–20) that goes beyond the Protocol to a stronger target of 20 per cent below the 1990 level by 2020. The scheme covers $CO_2$ emissions from large emitters from electricity generation, heat generation and energy-intensive industries. In total over 11 000 facilities in 31 countries are covered, corresponding to about half of the EU's $CO_2$ emissions and 45 per cent of its total $CO_2$-equivalent GHG emissions.[15]

In Phase 1 almost all emission permits were given to businesses free of charge. Phase 1 succeeded in establishing a price for carbon, free trade in emission permits across the EU and the necessary infrastructure for monitoring, reporting and verifying actual emissions from the businesses covered. In the first few months the spot price reached €30/tonne of $CO_2$ and, predictably, electricity utilities and some other businesses incorporated the carbon price in the prices of their products, such as electricity, making windfall profits totalling billions of euros. To make matters worse, governments made their national allocations of permits by guesswork under pressure from their energy-intensive industries. So, predictably, the total allocation of permits in the EU exceeded the total emissions from the sectors covered. When the market discovered this, the spot price of permits fell to zero.[16]

In Phase 2 the vast majority of permits were still issued free of charge. The European Commission tightened the cap slightly, but the Global Financial Crisis (GFC) that began in late 2008 reduced business activity, hence emissions and hence demand for permits, by an even greater margin. This led to a large and growing surplus of unused permits which held down the carbon price throughout the second trading period. At the end of Phase 2 it was unclear whether the EU ETS had been responsible for any significant reduction in emissions. A factsheet published by the European Commission shows a reduction per large emitter of about 17 kilotonnes or 8 per cent between 2005 and 2010,[17] but these reductions could have been caused mostly by the GFC and complementary policies that expanded RE and EE. According to a study by economists Stefano Clò and Emanuele Vendramin, in the electricity sector during the period 2005–11 the EU ETS promoted a small switch from coal to gas with emission reductions of 193 Mt.[18] If the EU ETS was indeed ineffective in Phases 1 and 2, the burden of meeting the EU's Kyoto Protocol target was transferred to the portion of the economy that was not covered by the ETS.

At the end of 2012, the EU ETS was in a coma. The following revisions to the scheme, that are in the design of Phase 3, may provide an intravenous drip. National allocations of permits have been replaced by a single EU-wide cap, which is designed to decrease by a small amount, 1.74 per cent per year, each year until 2020 and possibly beyond. In 2013 40 per cent of permits will be auctioned and this proportion will rise in the following years. However, there is growing concern in the EU that, to resuscitate the patient, stronger measures are needed. Options under consideration by the European Commission are:[19]

- increasing the EU's GHG target for 2020 to 30 per cent
- increasing the 1.74 per cent annual reduction in permits
- immediately withdrawing a large number of permits from the market (a longer-term option would be the institution of a carbon central bank to control the supply of permits in order to guarantee the stability of the carbon price within a predetermined fluctuation band)[20]
- including road transport in the ETS
- reducing access to overseas offsets

> • introducing a floor price, that is, a lower limit to the carbon price.
>
> These reforms would have to be passed by the European Parliament and by 74 per cent of the votes in the European Council, the representatives of the EU member countries. This will be difficult, especially while Poland, a major producer and user of coal, is opposed to changes.

## Carbon tax or ETS?

The principal advantages claimed for an ETS are that it specifies an emission target, that in theory it is more economically efficient, that it can be linked to ETSs in other countries and that it does not have the political stigma of being called a tax. Unfortunately, we have seen that in the real world of political power these apparent advantages may not be realised in practice. The European ETS, as it stood at the end of its Phase 2 in December 2012, could not on its own achieve the EU's GHG target. The tiny emission reductions attributed to it so far have to be weighed against the huge windfall profits given to electricity utilities at public expense (Box 8.2).

A carbon tax has several advantages over an ETS:[21]

- Taxes are a well-known instrument that can be readily implemented through the existing administrative system. They are generally less complicated, more transparent and less expensive to administer than emissions trading.
- A tax is generally preferred by the economic portfolios of government, because its effect on the budget is clear, at least in the short term.
- Revenues from taxes (and from emission trading with auctioning of permits) can fund compensation for low-income families that are vulnerable to higher energy prices, adjustment assistance for fossil fuel-producing communities and development of low-emission technologies and infrastructure.

- The carbon tax gives a much more stable carbon price than the volatile ETS, especially if the pathway for future increases in the carbon tax is set out in legislation. This gives more confidence to potential investors in low-carbon technologies and infrastructure.
- A carbon tax can be easily adjusted, within an initial broad pathway, to take into account other aspects of energy technology, such as security of supply, air and water pollution, water consumption and land degradation. It could also be expanded into an environmental tax.
- Unlike an ETS, a carbon tax does not allow speculators to create financial instruments that would provide money-making opportunities and hence increase the costs of the carbon price.

Carbon taxes have their own limitations, although these have straightforward solutions. Carbon taxes established in individual countries are not easily linked together into an international carbon tax. However, since there is at present little prospect for an international carbon price, it can be argued that it is better for individual countries to have an effective carbon tax than an ineffective ETS that could possibly be improved in the future. A carbon tax could be converted to an ETS if and when it becomes necessary. Another alleged limitation is that it is impossible to calculate objectively the level of the tax that would achieve environmental goals. The best one can do is to set out a future trajectory that gradually increases the tax until goals are achieved. To address this, the price trajectory would not be a line, but rather a pathway that allows minor adjustments to be made at intervals specified at the commencement of the scheme.

Both the carbon tax and ETS are regressive, in that they impact more on low-income earners than high-income earners. This disadvantage can be addressed by compensation: for instance, as part of its 'Clean Energy Future Plan', Australia lifted the income tax threshold substantially and the pension slightly while implementing a carbon price.

Value judgements are unavoidable in deciding whether to support a carbon price and in choosing between a carbon tax and a cap-and-trade ETS. My personal view is that a carbon price is necessary and should be implemented by all countries, provided low-income earners are compensated from the revenue obtained and provided there are complementary policies giving people alternatives to continuing on the business-as-usual pathway to environmental, social and economic disaster. Furthermore, the advantages of a carbon tax far outweigh its disadvantages and a carbon tax is more likely to be effective than an ETS designed by politicians. However, for both economic instruments it is essential that their respective designs are transparent and the potential outcomes of the various design features are open for public scrutiny and debate. If an ETS is chosen, it must have:

- a firm cap that is gradually tightened to drive deep cuts in emissions
- wide coverage (at least stationary energy, transport, non-energy industry emissions) and be applied upstream
- permits that are temporary licences to emit, not permanent property rights
- auctions as the principal means of allocating emission permits
- a transparent structure with low transaction costs
- compensation to low-income earners and to workers disadvantaged by the ETS
- no exemptions or compensation to GHG-polluting industries, apart from possible concessions to energy-intensive trade-exposed industries
- strict limits on the number of overseas offsets allowed.

Even with these characteristics it is unclear whether an ETS would be effective. Furthermore, whether the carbon price takes the form of a tax or an ETS, it can at best have an influence on the types of technologies and industries producing them, the $T$ factor in $I = PAT$, and on part of the $A$ term (consumption per person) through increased EE. However, to transform the economy to zero carbon we need to end the growth in overall consumption.

## Changing the economic system

At the time of writing, the GFC has imposed a pause in economic growth in the USA and much of Europe. The result has been widespread unemployment as high as 27.6 per cent of the workforce in Greece and 26.3 per cent in Spain in May 2013, the loss of savings and superannuation benefits by many, the imposition of severe austerity measures on social services and strong citizen resistance to these measures. What we are seeing, I suggest, are acute symptoms of a failed growth economy. Another symptom is the chronic failure of the economic system to shrink significantly the huge gap between the rich and the poor. A study by the Tax Justice Network finds that much of the financial wealth of the rich, estimated at $21–$32 trillion, has been invested in tax havens instead of trickling down to the poor.[22] To make matters worse, evidence suggests that wealth often trickles up from the poor to the rich.[23]

The most dangerous symptom is the failure of the economic system to protect the natural environment upon which we all depend. Nature provides photosynthesis that converts solar energy into a form that living things can ingest and utilise and oxygen in the air that we breathe. We also rely on nature to provide the great bio-geo-chemical cycles that enable us and the other species on Earth to exist and function: for example, the carbon cycle that, until recently, kept our climate in a balanced state; the cycle of the fluid of life, water; the cycle of nitrogen, an essential component of amino acids that form all proteins needed by living entities; and the cycle of the essential nutrient, phosphorus, that is a vital component of our bones and the molecule ATP that generates energy in our cells.[24]

All these free gifts from nature are under threat. One of the most urgent and serious threats is global climate change resulting from human-induced GHG emissions (see Introduction). A strong case can be made that the current economic system, built on the notion of endless economic growth on a finite planet, is one of the principal causes of this threat. In response to this threat the transdisciplinary field of ecological economics is developing. Some of its members are exploring the concept of a steady-state economy as the solution.[25]

A steady-state economy is defined as one that has a low throughput of energy and materials, with no growth in their use, and no population growth. Thus it is defined in biophysical terms, which is essential for environmental conservation. It does not prescribe a limit on growth in economic activity measured by GDP. In theory, some economic activities may not involve additional use of materials and energy – indeed, they could substitute for some energy-intensive and materials-intensive products and services. However, in practice, most GDP growth is associated with growth in the biophysical economy.[26]

Political economists, some ecological economists and those who work on sustainable development put the view that the new economy must be socially just, as well as ecologically sustainable. By social justice, they mean equal access to basic human needs, such as food and water, shelter, personal security, health services and education. One approach is to start with a strong definition of sustainable development that rejects trade-offs between ecological, social and economic aspects; here is my suggestion:

> Sustainable development comprises types of social and economic development that protect and enhance the natural environment and social equity.[27]

Another approach, by an alliance of ten leading scientists and other experts, is to define sustainable development in ecological terms, thus:

> Sustainable development ... [is] development that meets the needs of the present while safeguarding Earth's life-support

system, on which the welfare of current and future generations depends.[28]

and then embed this definition in a framework that includes the Millennium Development Goals, which focus on reducing extreme poverty in poor countries.[29]

A steady-state economy is not a failed growth economy. It is a new economic system designed to foster the above approaches to ecologically sustainable and socially just development, where development is not equated with economic growth. In the words of the New Economics Foundation, it is designed 'to share scarce planetary resources in ways that are just, sustainable and support the well-being of us all'.[30] Although the concept is in the early stages of development, macro-economic modelling by Peter Victor suggests that a steady-state economy could under some circumstances be compatible with low unemployment.[31] While several major research challenges still exist, such as how to maintain full employment during the transition to a steady-state economy, some of the broad policy requirements have already been set out by authors such as Herman Daly, Peter Victor, Tim Jackson, Rob Dietz and Dan O'Neil, cited above. These authors envisage a market economy operating within new constraints, incentives and disincentives. We can only list some of the proposed economic reform policies here:

- Internationally agreed caps on the mining of non-renewable resources and on the rates of extraction of renewable resources, to keep them below their rates of production.
- International caps on the disposal of toxic and hazardous wastes.
- Regulations, standards and monitoring to ensure all buildings and all consumer 'goods' are energy efficient, low in toxicity, durable and reusable or, failing that, recyclable.
- Environmental tax reform and a wide range of other policies to make environmentally damaging and unhealthy products and services more expensive and to hypothecate the tax revenue

raised to assist the development and commercialisation of environmentally benign and healthy substitutes.

- Tighter controls on advertising potentially harmful products.
- Private sector lending restricted to reserve requirements of 100 per cent as the default. However, reserve requirements could be adjusted according to the social and environmental benefits of projects, thus ensuring that money supply could only increase as real value is created.
- Public financial sector lending at low interest for large ecologically sustainable and socially just projects.
- A Tobin tax on international financial transfers, with the revenue hypothecated to the sustainable development of poor countries.
- GDP downgraded as an indicator of social and economic performance and replaced with a set of indicators that give a broad picture of the state of, and trends in, the environment, health, social justice, employment, income distribution and access to public facilities/services.
- Estate tax and land tax to target accumulated wealth.
- Guaranteed minimum income and cap on maximum income.
- Working time reduction.
- Law reforms to facilitate the formation and operation of cooperatives and other not-for-profit organisations.
- More research, including both macro-economic and biophysical modelling, of steady-state economic systems and possible transition pathways to those states.
- Constraints on the powers of large corporations in the interests of both democratic decision-making and fair markets.

While most of these proposed policies will be politically difficult to legislate and implement, the most difficult – and probably the most important – is the last one.

# Ending population growth without coercion

Different population policies are needed for rich and poor countries. In rich countries with very high per capita emissions, such as the USA and Australia, every additional person is on average responsible for much greater emissions over their lifetime than an additional person in a less developed country. Therefore, population growth must be ended as a matter of priority in the rich countries where it is still occurring.[32] In this case policy responses entail the removal of propaganda and financial incentives to encourage births and reduction in immigration quotas. Immigration into the rich countries has three components: the largest is usually business/professional, followed by family reunion and refugees, typically the smallest. From a humanitarian perspective, priority should be given to refugees and immigration should be capped by greatly reducing the business/professional component. This has the additional benefit of reducing the 'brain drain' from poor countries, which can ill afford to lose doctors, engineers and managers. Once a situation of Contraction and Convergence of GHG emissions has been achieved, with equal per capita emissions in all countries, then immigration quotas can be removed. However, policies to maintain the birth rate at replacement level will still be needed.

In less developed countries, the main policy needs for reducing births are the education of women, the provision of access to family planning services and economic development that provides social security for the aged.[33] To assist in achieving these goals, developed countries should increase their overseas aid budgets and loosen the strings that tie aid to their own businesses.

The principal barriers to population control policies are the Roman Catholic Church, which still opposes birth control, the property and housing industry, and indeed big business in general, which pushes for an excess of labour in order to keep wages down. Fearful of being unjustly accused of racism or being anti-refugees, many environmental non-government organisations have avoided the population issue or cloaked its discussion in global generalities.

# International agreements

Human-induced climate change is a global environmental, social and economic threat that must be tackled on a global scale as well as national and sub-national scales. Unfortunately progress on the international scale has been even slower than on the smaller scales. The United Nations Framework Convention on Climate Change (UNFCCC), which entered into force in 1994, has the worthy ultimate objective of 'stabilization of greenhouse gas concentrations in the atmosphere at a level that would prevent dangerous anthropogenic interference with the climate system'. The Convention adds that such a level should be achieved within a timeframe sufficient to allow ecosystems to adapt naturally to climate change; to ensure that food production is not threatened; and to enable economic development to proceed in a sustainable manner.[34]

Framework conventions are broad and contain value-laden words like 'dangerous' that are subject to interpretation. So in 1997 the Convention was strengthened by the Kyoto Protocol, which legally binds developed countries to (modest) emission reduction targets. The Protocol recognises that developed countries have contributed most to the concentration of GHGs in the atmosphere. The Protocol's first commitment period started in 2008 and ended in 2012. The second commitment period began in 2013 and will end in 2020. There are now 195 Parties to the Convention and 191 Parties to the Kyoto Protocol. The USA has not ratified the Protocol, nominally on the grounds that developing countries are not part of it, and several countries that participated in the first round withdrew from the second round (notably Canada, Japan, Russia and New Zealand).[35]

Has the Kyoto Protocol worked? It has been an important first step in global climate diplomacy and makes a strong implicit statement that the governments of hundreds of countries accept climate science. Since the Protocol was negotiated in 1997, global GHG emissions have surged 50 per cent to 2012, driven mainly by growth in China and several other rapidly developing countries who are not signatories. Emissions from ratifying countries have decreased, although most of this can be attributed to the collapse of GHG-emitting industries in Eastern Europe and more recently the GFC. Furthermore, much heavy industry migrated from developed to developing countries with the result that, when imports and exports are taken into account, there is a net increase in emissions in signatory countries.[36]

We need a successor to the Kyoto Protocol that places stronger targets on the rich countries, brings in the rapidly developing countries and recognises the needs of poor countries to develop their economies and social equity. This is a big ask! Not only do national governments face strong pressure from GHG-polluting industries, but many also don't wish to be among the first to act, believing that that may give short-term advantage to the economies of 'competing' countries. This situation is known as the Environmental Prisoners' Dilemma.[37] The best course of action for escaping the dilemma is for each nation to take urgent strong action, setting an example that encourages other nations to join.

A possible framework is the concept of Contraction and Convergence, alluded to in 'Ending population growth without coercion' above, in the context of limiting population growth. In this process the rich countries reduce their per capita emissions and the poor countries initially increase their per capita emissions until each country converges to the same safer level of average per capita emissions at some future date, such as 2040.[38] To avoid giving countries a perverse incentive to increase their populations, the population level taken to calculate the per capita emissions of each country should be chosen for a year *before* the commencement of the process. Since a safe level of emissions may have to be zero in the long run, it may be necessary to choose an initial convergence level of (say) 5 tonnes per person per year and then gradually bring this down to zero as sustainable energy and other measures become universal. Naturally there would be strong resistance to Contraction and Convergence by rich countries and rich elites in poor countries.

The earliest prospect for a renewed, stronger Kyoto Protocol, that covers all countries, is 2015. Meanwhile, those governments that are committed to effective climate action are focusing on the national and sub-national scales. Realistically, that is where the greatest potential lies for the near future.

## Conclusion

The mitigation of anthropogenic climate change and the principal, but partial, solution of a sustainable energy system face formidable barriers from the vested interests in greenhouse-intensive industries and other industries offering alleged solutions, such as nuclear power which is an expensive and dangerous diversion of limited effectiveness (see Ch 6). Another diversion is geoengineering, which would involve a new, risky, human modification of the climate as an attempt to alleviate the existing modification.[39]

To transform the energy system, a carbon price is necessary in order to increase the financial risk of investment in fossil fuels and related infrastructure and to send a signal that flows through the whole economy by making carbon-intensive products and services more expensive. Thus it internalises at least some of the environmental, health, social and economic costs of fossil fuels and so is consistent with and reinforces other policies to cut GHG emissions.

A carbon price is not sufficient, because the energy market has little resemblance to the perfect competitive market envisaged by neoclassical economists. Among market failures are the huge subsidies to fossil fuels and nuclear power, which far outweigh the small and declining subsidies to RE; price structures for residential electricity in many jurisdictions that do not reflect actual costs; split incentives for EE; the dominance of political decision-making for decisions to build infrastructure; the failure of a carbon price to take account of the many benefits of EE and RE additional to climate mitigation; and the failure of marginal economics to drive radical changes for the long term. Furthermore, policy failure in the design and implementation of a carbon pricing system, especially an ETS, can turn it into something very different from the idealised one considered by neoclassical economists to be highly efficient in economic terms. Hence we need a wide range of policy instruments in addition to a carbon price.

Apart from the specific policies needed to assist sustainable energy technologies to progress through the various stages of maturity (discussed in the next chapter), we must end population growth non-coercively and develop a new economic system that does not demand endless growth in the consumption of materials and energy on a finite planet and is far more effective than the existing system in fostering equal opportunity for all people in present and future generations.

Ultimately strong global targets are needed for reducing GHG emissions and for diffusing sustainable energy. However, in the absence of effective international treaties, action at the national and sub-national scales appears to offer the best prospects.

# 9: TARGETED POLICIES FOR RENEWABLE ENERGY

> To me, the path to a sustainable energy future seems very
> obvious … I think the technological challenge can be met …
> It remains to be seen whether the political and organisational
> challenges involved in recognising and addressing the need to
> change from the status quo can also be met on a reasonable
> timescale.
>
> Martin Green[1]

Have you seen the new drama with the following elements?

- Subsidies to the production and use of fossil fuels amounting possibly to a trillion dollars per year on a global scale.
- An emissions trading scheme that has transferred billions of euros from energy consumers to huge energy utilities without cutting emissions or supplying more electricity.
- An electricity industry whose business model is entering a death spiral, despite supportive institutions.
- Generous feed-in tariffs for renewable electricity replaced by retrospective charges following a change of government.
- Suburban neighbours bypassing electricity utilities by selling solar electricity to one another by cable.

Is this the plot of a thriller TV series by Danish national television? No, it's just some of the issues that must be addressed by governments by means of new climate action policies. We will expose them in this chapter.[2] But first, we must return to the policy thread.

In the previous chapter I presented a case that a carbon price is necessary, but not sufficient, for driving climate mitigation and, in particular, the transition to a sustainable energy system. Contrary to the beliefs of some neoclassical economists, for whom even the title of this chapter is heresy, a carbon price needs to be complemented by policies that facilitate the growth of specific technologies and the adoption of specific measures, including behavioural changes. In the realm of sustainable energy there are many different technologies[3] at different stages of maturity. Technologies at different stages of maturity require different mixes of supportive policies to assist them along the bumpy pathway towards becoming fully commercial and competitive with existing polluting technologies. This chapter first outlines the barriers to renewable energy (RE) technologies and then identifies appropriate policy mixes to assist the passage of RE technologies through each stage of maturity. Policies to support energy efficiency (EE) are discussed in Chapter 4 and policies for transport and urban form in Chapter 7. Since the science is telling us that the need is urgent, our principal focus will be on policies for disseminating technologies that are at advanced demonstration, early diffusion and commercially available stages. Subsidies to RE are discussed in the context of the much larger subsidies to fossil fuels and nuclear power.

## Barriers to technological maturity

In attempting to progress a technology through the various stages of maturity discussed in Chapter 2, a major challenge is the step between demonstration and early diffusion, often called the Valley of Death. Starting from a few medium-scale units and building a technology or system that can be manufactured on a limited scale is a very capital-intensive step. Venture capital is a risky business and its practitioners prefer to invest in products that are less expensive than energy systems and provide a rapid return if successful. At least one advantage of some RE technologies, such as new types of solar hot water or solar photovoltaics (PV), is that the size of a mature unit of hardware is small and so the commercialisation commitment is small compared with that for commercialising coal power with carbon capture and storage (CCS) or new nuclear power technologies. However, bringing engineered geothermal power stations and floating wind turbines from demonstration to early diffusion will not be cheap.

Once a technology reaches the early diffusion stage, it becomes 'bankable' and traditional sources of finance, such as debt financing and private equity can take over, *provided* there are government policies to bridge the gaps between the prices of the new and old technologies until the market has grown sufficiently to achieve parity. However, new technologies still face other barriers:

- Cultural barriers, such as the notion fostered by some neoclassical economists that a carbon price is sufficient for climate mitigation and the dissemination of sustainable energy systems.
- Institutional barriers, such as institutions whose business model is selling energy rather than energy services, electricity market rules designed for large centralised power stations and cumbersome administrative arrangements.
- Economic barriers, such as pricing systems that don't reflect the full costs of existing technologies and ignore subsidies to the production and use of fossil fuels. The absence of adequate infrastructure could be included here, although it is also a political barrier.
- Political barriers, notably well-funded campaigns by vested interests and their supporters against sustainable energy – a major barrier in the USA and Australia.
- Lack of education, training and information about sustainable energy.

Table 9.1 lists the principal portfolios of government policy options for lifting sustainable energy technologies over the barriers and through the various stages of maturity, provided of course that their

TABLE 9.1    **Government policy options for various stages of technological maturity**

| Stage | Policies |
|---|---|
| R&D | Research grants; tax deductions; support for patents |
| Demonstration | Capital grants; tax deductions; support for patents |
| Early diffusion | Feed-in tariffs; renewable energy certificates (RECs) (see 'Tradable renewable energy certificates' below); new infrastructure (eg, transmission lines; railways; cycleways); reverse auction (tendering); institutional changes; government procurement; education and training of workers and professionals; emission limits on conventional power stations; removal of subsidies for fossil fuels; regulations and standards for EE; standards for RE technologies; loan guarantees and other low-interest loans |
| Commercially mature | Increasing carbon price; feed-in tariffs or RECs to be phased out only after carbon price is consistently high enough to make them redundant; new infrastructure; continuing institutional changes; government purchases; education, training and information; time-of-use electricity pricing |

SOURCE The author, drawing upon Foxon et al. (2005).[4]

performance justifies support. The absence of a particular policy from a stage in this table does not mean that it is excluded, but just that it is less important than the others. For example, policies to foster research and development (R&D) may even be justified at the commercially mature stage where there can be strong competition to bring down prices. However, at that stage tax deductions are generally preferred over research grants and the dominant policies will be those that expand the market. Next we discuss some of the key specific policies needed at the early diffusion and commercially mature stages.

## Targets for sustainable energy

The scenarios described in Chapter 3 can assist us in forming practical visions of sustainable energy futures for the planet and our respective countries, states/provinces and local communities. Next we must answer the question: how can we get from here and now to those futures? One of the key early steps is to set challenging but achievable targets along the time-path to sustainability. The

European Union (EU) and California provide examples of an effective start with targets for 2020 and mechanisms to achieve them. Of course targets and mechanisms are also needed for 2030 and beyond.

The EU has committed to cutting its greenhouse gas (GHG) emissions by 20 per cent below the 1990 level by 2020. It has also offered to increase its emissions reduction to 30 per cent by 2020 if other major emitting countries in the developed and developing worlds commit to undertake their fair share of a global emissions reduction effort.[5] As part of the implementation of its greenhouse target it has legally binding targets for 2020 of a 20 per cent share of energy from renewable sources and a 20 per cent improvement in EE. Renewable Energy Directive 2009/28/EC sets mandatory national RE targets for achieving the 20 per cent share of RE in the final energy consumption and a 10 per cent share in transport, for the EU as a whole. Data indicate that the EU is on its trajectory towards the 2020 targets with an RE share of 12.7 per cent. However, progress among EU members is uneven and there are indications that the continuing Global Financial Crisis (GFC) has slowed investment. While PV is advancing ahead of schedule, off-shore wind is behind.[6] To progress towards their targets, many member countries of the EU use feed-in tariffs (see below); a few use tradable RE certificates (see below); levels of the carbon price in the EU ETS are too low in 2013 to be a significant driver.

In California the Global Warming Solutions Act (AB32), passed in 2006, mandates that GHG emissions be returned to the 2000 level by 2010, 1990 levels by 2020, and 80 per cent below 1990 levels by 2050. While AB32 is not directly related to RE generation, the state needs increased renewable capacity to meet these long-term emissions targets. Accordingly, the California Renewable Energy Resources Act (SBX1-2) of 2011 mandates 33 per cent renewable generation by 2020. To achieve this target California relies on a Renewable Portfolio Standard (see 'Tradable renewable energy certificates' below) applied to electricity retailers and publicly owned utilities.[7]

Targets are necessary but not sufficient. Without the development and implementation of policies to achieve the targets, they are merely futile aspirations. The previous chapter presented the case that a carbon price is necessary but not sufficient for the essential transition from fossil fuels to a sustainable energy system. One of the key policy options for specifically disseminating renewable sources of electricity at the early diffusion and commercially mature stages is the feed-in tariff.

## Feed-in tariffs

Feed-in tariffs (FiTs) are prices or 'tariffs' paid per kilowatt-hour (kWh) for electricity that is fed into the grid from designated RE systems. They are usually associated with laws that give priority of access of RE to the grid. The levels of the tariffs, which are set by governments, may vary with technology and geographic location and whether the RE technologies covered are competing with the wholesale or retail prices of electricity from the grid. They are usually guaranteed for a number of years. FiTs are intended to encourage the growth in the market for RE technologies by closing, or at least reducing, the gap in electricity prices between renewable and conventional energy and so give confidence to investors in RE. To this end FiTs were initially significantly greater than the prices of electricity purchased from the grid. However, as the size of the market increased, the prices of renewable electricity technologies declined and the levels of the tariffs were reduced, a process called 'degression'.

FiTs were originally introduced in California in 1984 under the name of Standard Offer Contract No 4. In 1991 they were introduced in Germany, where they have been very successful in driving the expansion of wind and solar PV. Until recently in Spain FiTs served to expand wind, solar PV and concentrated solar thermal (CST) with thermal storage. Nowadays FiTs of various types have spread to all the inhabited continents. Their recent status is extensively documented by RE industry analyst Paul Gipe.[8]

In most jurisdictions the cost of FiTs is covered by a small increase in electricity price paid by all or most consumers. In the few places where governments subsidised the scheme, the FiTs were suddenly reduced greatly or terminated when circumstances or the ruling political party changed. This was the case in Spain, where retroactive changes followed the advent of the GFC,[9] and in New South Wales following a change of state government.

There are two types of metering for FiTs, 'gross' and 'net'. For a gross FiT the tariff is paid on every kWh produced by a system connected to the grid, regardless of whether it is used by the owner or fed into the grid. The owner pays the retail price for any electricity they purchase from the grid. A 'net' FiT is paid on any surplus of renewable kWh fed into the grid after on-site consumption. If the value of the FiT is equal to the price of grid electricity at the site, the situation is called 'net metering' and is equivalent to having the RE input spin the meter backwards when generating.

A gross FiT has the advantages that the payment to be received can be estimated by the potential owner without knowing their consumption pattern and, for the same level of FiT, the owner's revenue is greater. If there is a net FiT in a jurisdiction, it may give little or no benefit to households whose occupants are at home for most of the day. Large-scale RE power stations compete with the wholesale price of electricity that is often quite cheap. Since they purchase little or no electricity from the grid, the distinction between gross and net FiTs is generally insignificant for them.

Advantages of FiTs compared with tradable renewable energy certificates (RECs) (discussed below) are:[10]

- reduced risk for RE developers entailing easier access to capital at lower interest rates
- low transaction costs
- easy entry for small and medium businesses
- policy flexibility – can be adapted as technology and markets change with time
- easy to make site- and technology-specific.

Disadvantages may arise if the tariffs are not adjusted to suit changing conditions and if investments are made in overpriced or poor performing projects.

## Tradable renewable energy certificates

This policy mechanism for building the market for RE technologies is generally associated with a national or state/provincial RE target and a quota or obligation placed on electricity suppliers, usually retailers, to produce a specified fraction of their electricity from renewable sources. Certified RE generators, including households, earn RECs for every megawatt-hour (MWh) they produce and can sell these along with their electricity to supply companies. To demonstrate their compliance with their regulatory obligations, supply companies then surrender the RECs to a regulatory body where they are extinguished. If the suppliers do not have sufficient certificates to cover their obligation, they must pay a pre-determined buy-out price for each missing certificate. RECs can be traded at market prices that differ from the official buy-out price. If there is more RE production than required by the obligation, the price of RECs falls below the buy-out price. If renewable and non-renewable generation prices became similar, the price of RECs could approach zero and there would be little or no subsidy to renewable generation. If there is less RE production than the obligation, the price of RECs increases above the buy-out price.

The mechanism was originally developed in several US states, where it is called Renewable Portfolio Standards (RPS). In Europe it is often called Tradable Green Certificates (TGCs), although the UK has chosen the term Renewables Obligation Certificates (ROCs). Australia has the Large-Scale Renewable Energy Target (LRET), which is the principal subsidy to wind farms, and the Small-Scale Renewable Energy Scheme (SRES), which has a fixed price for RECs and no target. The SRES predominantly supports domestic and small commercial solar PV systems. The introduction of the uncapped SRES avoids the potential limitation of certificate schemes that voluntary action by individuals and households does not add to GHG reductions, but simply reduces the task faced by electricity suppliers and industry within an overall RE target.

The strengths of the REC scheme are that, in theory, it delivers a predetermined and certain level of renewable electricity (RElec) generation and has an in-built incentive for electricity suppliers to source least-cost RElec. Its big disadvantages are that it has no in-built incentive to purchase electricity from a wide range of RElec types, gives less price certainty to potential investors than FiTs, and has no in-built structure that necessarily decreases the support given to RElec over time.[11] The former problem arises because, in the simplest form of the mechanism, 1 REC corresponds to 1 MWh of electricity generated from any RE technology included in the scheme. Generators wishing to acquire their obligation at least cost naturally seek the cheapest certificates, which are usually wind, hydro and landfill gas. Large-scale solar electricity, both PV and CST, is too expensive to receive much benefit from RECs. Now that the price of electricity from PV has fallen substantially, RECs can assist them to a limited degree in some jurisdictions. Nevertheless, the simple form of a RECs scheme generally fails to drive the broad base of RE technologies needed for an electricity supply system that will transition to a predominantly renewable system. Different RE technologies have different variations by time of day, season and year. Their resources tend to be in different geographic regions, with solar more prevalent at lower latitudes and wind at higher latitudes. So a broad mix of RE technologies has the least variability and the least need for flexible back-up and storage. This can be achieved best either with FiTs or a complex RECs system in which each eligible technology has its own target.

With or without this refinement the theoretical benefit of RECs schemes can be lost by poor design. For instance, in the Australian scheme the price of RECs has been low for several years because a large excess of RECs was created by including solar hot water as an eligible technology and also 'phantom' RECs were created by issuing initially 5 RECs per MWh of solar PV.[12] The majority of studies comparing RECs and FiTs find FiTs to be more effective.[13]

## Removing subsidies to fossil fuels and nuclear energy

Huge subsidies to fossil fuels, and in some countries nuclear energy, are slowing the diffusion of sustainable energy technologies. Nancy Pfund, a Managing Partner of venture capital firm DBL Investors, and graduate student Ben Healey have reviewed US federal government subsidies 'within the rich historical context of US energy transitions'. They found that, over the first 15 years of each subsidy's life, in inflation-adjusted 2010 dollars, annual subsidies to oil and gas averaged $1.8 billion, nuclear subsidies averaged $3.3 billion and RE subsidies averaged less than $0.4 billion.[14] However, unlike the subsidies to RE, fossil fuel and nuclear subsidies have existed for many decades and are not as yet decreasing.

An energy or transport subsidy exists where government action or inaction lowers the cost of production, raises prices received by producers, lowers prices paid by consumers or prevents full cost recovery for a service. Some subsidies involve direct payments from governments to businesses, but many subsidies are created indirectly through tax rules and government practices. Typical types of subsidies provided by governments include:[15]

- direct payments and rebates
- favourable tax treatment
- provision of infrastructure and public agency services below cost
- public contributions for R&D
- provision of capital at less than market rates
- failure of government-owned entities to achieve normal rates of return
- utilities charging electricity users for poor utility investments and
- trade policies, such as import and export tariffs and non-tariff barriers.

To these economic/financial subsidies must be added the subsidies resulting from a failure to internalise the environmental
and health costs of fossil fuels. Carbon pricing or, more generally,
environmental tax reform, is one way of doing this. Although
there will always be controversy about the appropriate level of
the price or tax, existing studies on the externalities of energy use
generally find that the additional costs range from being comparable to the existing price to being much greater. That is, the
true cost including externalities is up to several times greater than
the price excluding externalities.[16]

Without addressing external costs, the International Energy
Agency (IEA) estimates that partial global subsidies to fossil fuels
in 2011 amounted to $523 billion per year, six times the subsidies
to RE. The IEA interprets the 30 per cent increase in these subsidies from 2010 as reflecting 'higher international energy prices and
rising consumption of subsidised fuels'.[17]

The IEA estimate is limited to just one of the types of subsidies
covered by the definition at the beginning of this section, namely
subsidies that reduce consumer prices 'below those that would prevail in an open and competitive market'. Total subsidies are much
higher. This is demonstrated by the IEA result that Australia has
zero subsidies, while a detailed analysis by researcher Chris Riedy
identifies annual subsidies to the production and use of fossil fuels
amounting to at least $10 billion.[18]

The IEA finds that the largest proportion of fossil fuel subsidies is for the production and use of petroleum products. Among
consumers by far the highest proportion of the subsidies flows
to high-income earners. However, modest benefits also flow to
low-income earners in several poor countries, for instance, from
subsidies on kerosene burned to give dim, polluting lighting for
village people in India and Bangladesh. To avoid further impover-
ishing these people when subsidies are removed, additional policies
are needed to facilitate the replacement of kerosene with solar
lighting, which is bright and non-polluting. In the absence of kerosene subsidies, solar lighting with batteries would be a lower cost

option in many rural villages if it could be paid for in instalments or leased.[19]

More generally, the funds saved by the removal of subsidies can be returned to the community to facilitate the transition to a sustainable energy system and to assist workers and households disadvantaged by the transition.

## Innovative finance for sustainable energy

RE and EE projects typically have high capital costs and very low operating costs. So, in initiating a project, a key challenge to be overcome is obtaining finance. For government programs revenues from carbon pricing and the removal of subsidies to fossil fuels are potential sources of finance. While governments can stimulate investment, the major share will have to be taken by private companies and investors. As a result of the ongoing GFC, banks are reducing their allocations for loans to individual projects, which are seen as more risky financially than other options such as increasing equity in successful projects. Even where sufficient project finance is available, a global or national strategy based on applying for individual project loans is a slow process from the viewpoint of a strategy for urgent climate mitigation. Indeed, the IEA has estimated, in its 450 Scenario, that cumulative investments amounting to $16 trillion above its New Policies Scenario[20] must be invested worldwide over the period 2012–35. The major part of this would be for EE (from reduced electricity demand, buildings and industry) and RE. In the 450 Scenario EE would contribute nearly three-quarters of the GHG reductions and RE 15 per cent in 2020; in 2035 EE would provide 45 per cent and RE 23 per cent.[21]

Providers of venture capital generally prefer to invest in projects that can give rapid repayment and do not have such large up-front costs as RE, such as information technology. The Australian Labor government has attempted to stimulate the venture capital market for sustainable energy by introducing a government provider, the Clean Energy Finance Corporation.[22] This too is a slow process and subject to the whims of new governments. (Indeed the Coalition government, elected in September 2013, has announced plans to scrap the Corporation.)

An alternative financial instrument is the 'climate bond', sometimes called 'green bond'. Climate bonds enable long-life projects to be pooled so that they are of interest to big investors. In principle they could provide trillions of dollars of finance for RE and large-scale EE programs.

### Climate bonds for pooling large-scale projects

A bond is similar in some ways to a term deposit. Investors receive certificates that state how much the bond is worth, the fixed amount of interest that will be paid on specified dates (usually semi-annually) and the date of maturity, at which point the face value of the bond is returned to the bondholder. Bonds are intended especially for big investors such as superannuation or pension funds, insurance funds, sovereign wealth funds and large businesses. Unlike term deposits, bonds can be traded. In general a bond pays low interest; however, it has low risk. Although a bond is not a new entity, its large-scale application to the sustainable energy transition would be new. Climate bonds need certified standards, to ensure that funds are really directed at low-carbon projects. In addition, a management strategy must be adopted that secures a low risk classification by credit agencies.

The Climate Bonds Initiative is an investor-focused not-for-profit network, promoting large-scale investment in the low-carbon economy. It promotes safe and secure investments suitable to the needs of pension and insurance funds. Its goal is to rapidly help develop a large-scale Climate Bond sellers' market and a large-scale Climate Bond buyers' market and then assist in connecting them. Its Board has created standards for wind energy bonds and has certified its first bond, soon to be launched, as 'low carbon'. Solar and EE investments will be the next to be certified.[23]

A study by the Climate Bonds Initiative found[24] that in 2013 'the climate-themed bond market is not niche, lacking scale or liquidity'. Bonds outstanding are worth $346 billion in total, of which 89 per cent is 'investment grade'. The investments considered are in seven theme areas. In order of their contributions to the total they are:

- rail transport, excluding coal railways
- energy: RE, excluding large hydro in tropical regions, but (controversially) including nuclear
- climate finance
- buildings and industry, focused on EE
- waste and pollution control, mostly industrial recycling
- climate-resilient agriculture (none found yet) and certified 'sustainable' forestry
- water supply resilient to climate impacts (none found in this study).

The Climate Bonds Initiative points out that ratings agencies are still over-estimating the policy and resource risks of RE, while under-estimating the carbon penalty risk of fossil fuel projects. Therefore, to build investor confidence and hence expand the climate bond market, government policies to support climate bonds – such as tax deductions, loan guarantees, FiTs and tradable certificate schemes – are vital for their effectiveness.

### Pooling small investments for medium-scale projects

Medium-scale RE projects generally face more complex regulatory arrangements than small-scale residential projects, thus making them more expensive per unit of energy generated. In addition, especially for medium-scale wind power, fixed costs such as grid connection are a much larger fraction of total costs than for large-scale projects. So it is often harder to obtain finance from commercial lenders for medium-scale projects. However, community-initiated projects have shown the way forward for medium-scale RE in Denmark and Germany, as discussed in

Chapter 10, 'Action by cooperatives and other community groups'.

An innovative approach to financing medium-scale projects has been initiated recently by US on-line company Mosaic, formerly Solar Mosaic, which provides a platform to coordinate investments from $25 upwards in solar PV projects. It offers a low-risk investment with a good rate of return, typically 4.5 per cent per year, over a fixed term. In May 2013 the company announced that 823 people had invested a total of over $1 million in its largest project, a 487 kW installation on New Jersey's Wildwoods boardwalk.[25]

### Leasing for small-scale projects

Leasing is another innovative option for financing renewable electricity, especially residential PV. A solar company called Sungevity leases residential solar PV in the USA and Australia. Using satellite imagery of the home, Sungevity gives an on-line quote. There is no up-front fee and the monthly fee covers installation, rental of the system, monitoring, maintenance, repair and a guarantee of performance.[26]

## Transmission infrastructure

Like fossil fuels, RE resources are not distributed equitably over the world or even over individual countries. New transmission lines are needed to bring renewable electricity from wind-rich and solar-rich regions to concentrations of consumers and industry. This is what was done when the existing fleet of older coal-fired power stations was constructed. The transmission problem actually predates the rise of RE. In Europe there are inadequate transmission links between several countries, limiting electricity trading across the continent. In Japan, recovery from the tsunami and Fukushima nuclear disaster is impeded by incompatible transmission zones within the country.

Now that RE is growing rapidly to become a significant proportion of electricity generation in several regions, the need for government policies to upgrade old lines and build new ones has become acute. China has installed wind farms at such a high rate that many of them are still lacking connections to transmission lines.[27] The states of Texas (USA) and South Australia have huge wind potential that cannot be tapped until transmission links to neighbouring states are augmented. The urgent policy need is for new transmission 'spines' within countries. As well as tapping large RE resources, the benefits of long-distance transmission links include greater energy security, enhancing trade and competition, smoothing of fluctuations in RE through geographic diversity, and temporal smoothing by connecting different time zones.

There are several proposals to build high-voltage undersea transmission circuits off-shore to link wind farms, solar power stations and hydro across nations or states. Off-shore links have the additional advantage of bypassing on-shore transmission bottlenecks and community resistance. The European Wind Energy Association has examined scenarios for building an off-shore grid in the North and Baltic seas[28] and such developments are being actively pursued by the electricity industry. In the USA a consortium called Atlantic Wind Connection has proposed to build a high-voltage DC transmission circuit off the east coast to link future off-shore wind farms from Maryland to New Jersey to the on-shore grid.[29] A longer-term intercontinental project is the Desertec proposal to link solar power stations and wind farms in North Africa to Europe.[30] This project has political obstacles and has recently suffered a setback with the withdrawal of two of the major consortium members, Siemens and Bosch.

## Policies for renewable heating

In the northern regions of Europe, Asia and North America heating comprises a large fraction of energy use and GHG emissions. Despite long-established, very efficient combined heat and power (CHP) stations providing district heating, policies to support RE sources of heating are less advanced in terms of implementation than for renewable electricity. As in the case of renewable electricity, policies should be chosen to be appropriate to the stage of maturity of a technology, as shown in Table 9.1. The main differences are that it is more complicated to apply RECs to heating and impossible to apply FiTs in the absence of a grid for transmitting heat. However, bonus schemes, in which a plant operator receives a bonus per megajoule on top of the normal rate, may be feasible for production of heat for local use. Britain has recently introduced a Renewable Heat Incentive for biomass boilers and air-source and ground-source heat pumps that heat buildings.[31]

## New business models for electricity industry

As EE, energy conservation and renewable electricity are growing, existing business models of the electricity industry are collapsing. The 'disruptive' sustainable energy technologies and behaviour are challenging existing industry structures – both traditional vertically integrated utility and market-based arrangements – at both the wholesale generation and retail market stages and at transmission and distribution.

### Merit order effect

At the wholesale generation level of a restructured electricity industry, different power stations make offers at each time-step to supply electricity into the market. In theory each participant is incentivised to offer at its short-run marginal cost of generation, that is, its fuel cost plus other operating costs, although in practice it does not always make this supposedly economically 'rational' decision. Then the market operator ranks power stations in order from the lowest to the highest priced offer. This order is known as the 'merit order' and the power station with the lowest offer is called the 'top of the merit order' and has the first priority for dispatch. At each time-step the market operator dispatches power stations in merit order, subject to a range of security and other constraints, until demand is met and, at least in theory, electricity is supplied at the lowest cost. The offer of the last power station to be dispatched determines the price paid to all the generators dispatched at that time-step.

Incidentally, if the bidding occurs near a peak in demand or a period when a significant fraction of supply is not available to generate, one of the 'power stations' to be dispatched could in principle be a virtual one comprising a demand reduction from an electricity-intensive industry or community group. If organised in advance, demand reduction could slice the top off the peaks in demand or help fill gaps in supply.

Wind farms, solar power stations and most other types of RE power stations have lower operating costs than conventional power stations that burn or fission a fuel. Hence these renewable stations are ranked higher in the merit order for dispatch into the grid. Thus, as renewable capacity increases, conventional base-load power stations, such as coal and nuclear, are operated less frequently. Their capacity factors decrease and they are displaced to the role of intermediate load. Thus it becomes increasingly difficult to pay off their high capital costs and some may have to be closed down. However, the economic problem for conventional base-load generators goes beyond reduced operational periods in general: in particular, they generate less during peak periods, when all operating power stations receive the highest prices for electricity. These are often the periods when they used to receive the most income.

To make matters worse, thermal power stations designed for base-load operation are generally technically unsuited to rapid changes in output due to thermal stresses and high minimum operating levels. Therefore they cannot respond to rapidly changing output requirements. To remain on-line in an electricity market, they have to bid prices below their marginal cost thus exacerbating the challenge to their economic viability.

The rise of RE power stations, mainly wind farms so far, and the associated displacement of fossil-fuel power stations is reducing the wholesale spot price of electricity. This price reduction, associated with the displacement of conventional power stations, is known as the 'merit order effect'. The growth of solar PV at the residential stage also contributes to the merit order effect by reducing the daytime parts of the peak demands and hence the price paid to generators during those periods. The merit order effect has already been observed and/or modelled in Denmark,[32] Germany,[33] Spain,[34] a group of European countries,[35] Texas[36] and South Australia.[37]

What are the policy implications of the merit order effect? From the viewpoint of sustainable energy the displacement of conventional base-load power stations is a necessary part of the transition to a sustainable energy future. Shareholders in the conventional system will suffer financial losses as a result of stranded assets whose value will inevitably decrease. Only the rapid retrofitting of CCS to these polluting power stations could avoid this outcome. Given that CCS from large coal-fired power stations has not even entered the demonstration stage of maturity, such retrofits are very unlikely to occur on a significant scale within the next 15–20 years. If and when they do occur, they may not be able to compete economically with RE, except perhaps in a few locations with convenient geological structures or facilities nearby for CCS.[38]

In the view of environmentalists and many economists, governments should resist any claims for compensation by fossil-fuelled generators. The industry and its investors have had ample warning that carbon pollution must be paid for and polluting technologies must be phased out. Kodak was not compensated for the displacement of film photography by digital and fixed-line phone manufacturers were not compensated for the rise of mobile phones. Resistance to claims for compensation is the key policy implication of the merit order effect.

While the merit order effect is currently working to the advantage of RE, in the longer term the effect will become a challenge to investment in all technologies that can contribute to supplying electricity demand, including RE. This is because a reduced wholesale price of electricity increases the payback period on the capital cost of the technology, thus discouraging investment in new capital-intensive power stations. One solution under discussion in Germany and Australia, and implemented in many restructured electricity industries, is to pay power stations for supplying generating capacity as well as energy. However, to ensure reliability of electricity supply with high RE penetrations, it would be necessary to direct such payments to flexible, fast-response plants such as bio-fuelled gas turbines, hydro with storage dams, CST with thermal storage and other storage technologies that can deliver high power upon demand at short notice, say 10 minutes or less. Variable RE generation facilities, such as wind farms and solar farms without storage, could also be paid lower rates appropriate to their lower 'capacity credit'.[39]

Although purchasers of wholesale electricity, including electricity distributors and retailers, are benefitting from reduced prices resulting from the merit order effect, many retailers are not passing on the reductions to their customers. This may be because retailers and distributors are also experiencing a threat to their business model from renewable electricity, the so-called 'death spiral'.

### Death spiral

The IEA is projecting in its New Policies Scenario to 2035, that 'Energy demand barely rises in OECD [Organisation for Economic Co-operation and Development] countries, although there is a pronounced shift away from oil, coal (and, in some countries, nuclear) towards natural gas and renewables.'[40] At the retail level of the electricity supply industry, demand for electricity from the grid is already decreasing or ceasing to grow in several parts of the industrialised world, including the USA[41] and Australia.[42] This decrease has a combination of causes which are difficult to separate: EE programs; the growth of solar PV in the residential and commercial sectors as its price decreases; a slowdown in economic activity; and the increasing price of retail electricity which is driven predominantly by the growth in residential air conditioning plus small components from carbon prices and subsidies to RE where they exist.

The retailer often bills the customer for two components of supply: a fixed charge for 'supply' or 'service' and a variable charge that increases with the quantity of electricity purchased by the customer. The fixed charge may be either a lump sum item on the bill or be spread across a first tranche of energy use, or across all energy used. The proportions of the various price components vary by location and in some jurisdictions by time of day as a result of the market price varying with demand. The breakdown of the variable price paid by retail customers in the Australian Capital Territory (ACT) into its components is illustrated in Table 9.2.[43] Unlike California, the ACT has a flat tariff option for small customers, that is, one that does not vary with time of use.

Declining demand for electricity poses a serious problem for the electricity retailer and the distributor (network operator). If

TABLE 9.2 **Components of retail tariff in the ACT, 2010–11**

| Item | Cost in $/MWh |
|------|---------------|
| Electricity purchase | 58.57 |
| Energy trading desk operation | 0.76 |
| Environmental compliance costs, eg, Mandatory Renewable Energy Target | 5.15 |
| National Electricity Market fees | 0.76 |
| Energy losses (%) 5.92 | |
| Total energy purchase costs | 69.01 |
| Network costs | 71.44 |
| Total retail + energy + network cost | 151.01 |
| Retail margin (% of sales) 5.40 | |
| Total retail price | 159.16 |

SOURCE Independent Competition & Regulatory Commission (2011).
NOTE There is also a fixed 'supply' charge of 73 cents per day.

they raise the price per unit of electricity to compensate for the smaller number of units sold, it gives further incentive for customers to install EE technologies and solar PV, which in turn reduces the demand for grid electricity, which triggers another tariff increase, and so on. This pathway, which leads to the collapse of the business model of the retailer and network operator, is called the 'death spiral'.[44]

To survive, at least for the time being, the network operator and retailer may choose the alternative pathway of increasing the fixed charge to the customer. This has been the response of fixed-line telephone companies to the rise of the mobile phone. This increase has accelerated the shift to mobile phones for residential and small business customers, another death spiral. However, the transition is slower for retail electricity customers, since at present very few of them can afford to install a large enough battery bank to provide electricity through the night and, even more challenging, to achieve supply reliability as high as the grid (eg, a week of cloudy weather). Without this level of reliability it seems unlikely that many customers of grid electricity would disconnect from the grid. They have little choice but to stick with a retailer and pay the fixed charge until battery prices, which are declining quite slowly, reach a level where disconnection and independence make economic sense. Setting a high fixed charge also has the adverse environmental effect of removing the economic incentive for households, small businesses and other customers of electricity retailers to reduce their electricity demand. If you have to pay essentially the same bill no matter how much or how little electricity you use, you have little incentive for EE and energy conservation.

Another response by electricity retailers and network operators is to lobby governments for the removal of subsidies to RE. As a result, FiTs and RECs schemes have been cut in several countries and states/provinces much more severely than through planned degression.[45] Some retailers and distributors are going further by pressuring governments to impose retrospective reductions in FiTs and even additional charges on owners of renewable electricity systems located on the customer side of the meter. Such moves have been attempted by three Australian state governments, but each time the growing political strength of PV owners and the RE industry has successfully resisted these proposals. The electricity industry claims that solar PV owners are increasing the use of the poles and wires, to the extent that upgrades are needed in some cases, without paying for them. The owners and RE industry counter by pointing out that network costs are included in the electricity tariff, as for example shown in Table 9.2, and that the feed-in of RE confers benefits that are so far unquantified.[46]

RE owners and industry add that the current increases in the costs of poles and wires and capital expenditure on additional peak-load power stations are due in some jurisdictions to the increasing use of air conditioners, some of which are cheap and very inefficient in their use of electricity, and only a small degree to subsidies to RE. A 2012 Australian Government Energy White Paper states: 'For example, while it may cost around $1500 to buy and install a 2-kilowatt (electrical) reverse-cycle air conditioner, such a unit could impose capital costs on the energy system as a whole of $7000 when adding to peak demand.'[47] Therefore, I recommend that these costs should be recovered from a demand charge proportional to the maximum demand that the customer chooses.

### A sustainable electricity system scenario

The current transitional conflict between the electricity industry on one hand and retail customers and RE owners on the other, is caused to a large degree by the resistance of the industry to change. Industry incumbents generally wish to continue with the same industry structure and business models as in the past, despite the fact that circumstances – and in particular the business environment – have changed. Maintaining past institutions is clearly impossible. To develop solutions we need a plausible scenario for a future electricity industry and policies enabling that scenario to be achieved. The ultimate goal must be a supply–demand system that has no or very low GHG emissions, is safe, has secure energy sources for the long term, is as reliable as the present system and is affordable, although not necessarily as cheap as the existing system. Here is an incomplete scenario.

In 2035 (say) there is still a grid of long-distance transmission lines and local distribution lines. The grid and the electricity market (where the market is retained) have changed their structure and operational practices reflecting a predominantly RE supply at both wholesale and retail levels. New high-voltage transmission links join regions with high wind and solar resources to the demand centres. To balance supply and demand and to keep the frequency constant, the large-scale part of the system has much more storage than in 2013. This storage takes the form of pumped hydro, biofuelled gas turbines, CST with thermal storage and large flow batteries, and is located on the transmission and distribution grids as well as at RE generation sites. There is a much greater role for geographically distributed local generation for local use where small-scale technologies supply residential consumers and medium-scale technologies supply community groups, the commercial sector and small to medium-sized industries. In towns and cities large industrial users and high-rise buildings buy electricity from the grid, supplemented by on-site or near-site solar PV wherever possible. In locations remote from the grid large industries use mostly RE where the resource is good. Some residential electricity users have solar PV with battery storage and are disconnected from the grid. Others remain connected and use the battery storage in their electric vehicles to sell excess energy and power to the grid at times when the price is high. Others sell excess electricity directly to their neighbours using their own low-voltage, low-capacity cables, in competition with the grid.[48]

Virtual power stations, based on the coordinated supply from thousands of geographically distributed residential PV systems, supply blocks of power that are less variable than from a large solar PV station located at a single site. Unlike concentrated solar PV and CST power stations, this dispersed coordinated 'power station' with flat-plate solar collectors still generates during most overcast conditions, albeit at a reduced level.

Another type of virtual 'power station' supplies 'negawatts' (demand reduction) into the grid. It coordinates the offloading of large blocks of electricity demand at critical times, playing an important role at the wholesale level in shaving peaks in demand and covering gaps in supply. Such stations also operate to reduce excessive demand on distribution lines and nodes when required. Smart devices, attached to the power cords of air conditioners, refrigerators and other appliances, are pre-programmed to switch off the appliance for a period of time when the electricity price reaches a predetermined level. These devices may be programmed by consumers and, given appropriate contracts, the negawatt provider or the electricity distributor/retailer.

In this 2035 system the business roles of distributors and retailers are rather different from their roles in 2013. EE expert Alan Pears envisages that some of the distributors that have survived the transition have regional electricity storages to store low-cost electricity from generators and sell it at premium prices when it is really needed. If the storages are located strategically they can also store exports of distributed small-scale renewable electricity into the distribution network, reducing the need for expensive augmentation of the network. Along with other businesses, distributors (or subsidiaries thereof) could also seek a licence to bid demand reduction into the wholesale electricity market.[49]

However, in 2035 distributors are no longer regional monopolies. They and retailers (if they exist) have real competition. Neighbours are selling renewable electricity to one another. Appliance manufacturers are offering to households and businesses integrated building energy systems that include on-site generation, storage, smart controls and smart, efficient appliances. 'These companies understand that economy of scale doesn't have to come from bigger systems: mass production also works.'[50]

When asked about the electricity pricing system in 2035, I find that some sections of my crystal ball are cloudy. There appears to be some kind of a market that is freer in some respects, but more planned in others. However, the ball does reveal that the industry's old business model collapsed around 2020 and that electricity retailers have been replaced by businesses offering EE, local generation, generation sharing on the customer side of the meter, and demand reduction strategies. These are similar to the energy service companies (ESCOs) discussed in Chapter 4, 'Overcoming barriers: strategies and policies'. Since the vast majority of the costs of the predominantly RE supply system are fixed capital costs, customers in 2035 are paying a capacity charge proportional to a contracted maximum demand. However, the fixed charge paid by the small consumer is insufficient to cover the fixed costs faced by the supplier. Therefore, the variable component of the bill for RE, that is, the charge for each unit of electricity used, has been increased to cover that shortfall. It also contains a large carbon price and FiTs for RE. These tariffs take into account any additional costs imposed on the network and the benefits given to the network by RE. Like electricity purchases from the grid, these FiTs vary by location and the status of supply and demand. One thing is clear: once the difficult transition has been made, electricity prices in a 100 per cent renewable electricity system should become quite stable.

## Policies for transforming the electricity industry

The 2035 scenario suggests a new kind of market for electricity, and more generally for energy services. Suggested policies to enable such a future include in general terms:

- Existing policies that subsidise fossil fuels, bias market rules towards large centralised power stations and give monopoly power to distributors, should be terminated.
- Legislation and market rules should be implemented to facilitate the creation and operation of virtual power stations that can coordinate many electricity consumers or small-scale renewable electricity generators and bid demand reduction or RE into the wholesale electricity market or sell to distributors.

- There must be no compensation to fossil-fuelled generators with stranded assets.
- Electricity customers connected to the grid could be billed separate demand and energy rates. The demand rate could involve a contracted maximum demand received by the customer and a contracted maximum feed-in of RE from the customer to the grid. The energy rate would depend on time of use. (Needless to say, the advantages and disadvantages of this billing option and alternatives need further research and debate.)
- RE generators, fast-response dispatchable power generators (including biofuelled gas turbines) and stored energy suppliers should be paid for the availability of their capacity and separately for energy supplied to the market.
- Policies should be implemented to facilitate the financing and approval of medium-scale RE systems by local communities, commercial businesses and small industries, to transfer electricity at low voltage between users within the same site or neighbourhood, and to set standards and regulations for such transfers.

Since we are still at the early stage of debating the future of the electricity industry while it is undergoing rapid change, the policies suggested above should be considered as an input to further discussion rather than a definitive plan.

## Conclusion

The transition to an energy system based predominantly on renewable sources needs multiple policies, depending on the stages of maturity of the technologies. It presents significant institutional and operational challenges.

Because of the need for urgent climate action, this chapter focuses on expanding the market for technologies in the early diffusion and commercially mature stages. Targets and a timetable are essential. As emphasised by the IEA, it is also essential to remove the subsidies to the production and use of fossil fuels.

To support the expansion of the market for renewable electricity generation technologies in these stages, FiTs have been the most successful policies, while tradable RECs schemes have provided good support to the lowest-cost RE technologies, especially wind. Large-scale renewable electricity also needs an augmented electricity grid. Although there is little experience as yet with policies to expand renewable heat, bonus schemes may be an effective option for the above stages. Also needed to finance renewable electricity, renewable heat and EE are innovative instruments, such as climate bonds for pooled investments in many large-scale projects, coordinated small investments to fund medium-scale projects and leasing for small-scale projects.

The rapid growth of renewable electricity in some regions is challenging the business models of the electricity industry at both the wholesale and retail stages. At the wholesale or generation level, high penetration of wind power, supplemented by solar PV at the retail level, is reducing the price of electricity and substituting for base-load coal-fired power stations. This merit order effect is just what is needed. Governments should resist claims for compensation for these predictable stranded assets.

At the distribution/retail level, electricity demand has reached a plateau in some developed countries and is declining in others, as a result of expanding solar PV, EE, energy conservation and the ongoing financial crisis. As a reaction, retailers are increasing prices per unit of electricity and fixed charges, a sure recipe for entering a 'death spiral' for their businesses. Some are lobbying governments to slow the growth of PV. Some preliminary suggestions are offered in this chapter for moving towards a new business model as the transition to sustainable energy continues.

To overcome the barriers and redirect society and the economy to a rapid transition to sustainable energy, we need a large, well-informed, committed, social movement to exert countervailing pressure on decision-makers in government, business, trade unions, professional organisations and the wider community to implement effective policies. Building the social movement is discussed in Chapter 11.

# 10: WHO WILL DRIVE THE TRANSITION?

Alone we can do so little. Together we can do so much.

Helen Keller[1]

Given that we need urgent, rapid and effective action to transition away from fossil fuels to an ecologically sustainable energy system, based on the efficient use of renewable energy (RE), which sectors of society will be the prime movers for change?

In most countries governments have the greatest potential to take on this role. Within the nations they govern they can establish energy efficiency (EE) standards for buildings and appliances, set national targets for EE and RE, legislate pollution controls, purchase 'goods', tax 'bads', implement a carbon price, set up tradable emissions permits or tradable RE certificates, low-interest loans and loan guarantees, and restructure the electricity industry. They can fund research, development, demonstration, infrastructure, education and information. Externally governments can deal with the global problem of climate change by making firm international agreements to cut emissions, assisting the poor countries to adapt to climate impacts and diffusing sustainable energy technologies and practices across the globe. Unfortunately even less has been achieved to date at the international level than at the domestic level of many countries (see Ch 8, 'International agreements').

In practice, some governments are constrained by close ties to the fossil fuel-producing industries, very big energy users such as aluminium and steel, and trade unions of workers in these industries. This is a major barrier to change in the USA, Poland and Australia, among other countries. Vested interests exert pressure on governments by lobbying, backed up by advertising, by influencing the commercial mass media and by establishing close relationships with government officials responsible for regulating the industries with vested interests.[2]

Big business has the capacity to change direction even more rapidly than governments. This is to be expected, since it does not have democratic decision-making and is generally regulated lightly. However, its legal responsibility is to its shareholders, not to the public interest. As long as it has sufficient customers for its products and services to make a profit, it does not need to change. Nowadays many large businesses and industries claim to be implementing the concept of 'green business', but in the vast majority of cases their actions are small token gestures that do not have significant impact on their core polluting activities.[3] Thus electricity utilities whose principal fuel is coal spread pictures of wind farms and solar power stations across their homepages.

As a reaction to this, some supporters of climate action have argued that actions by individuals and households can substitute for actions by unwilling governments. Is this true? This chapter discusses the potential roles of individuals/households, communities and big business in the great energy transition.

## Limits to individual action

While individual/household action has a role to play in social change, its effectiveness is often exaggerated with simplistic arguments. One such argument is to point out the obvious truth that all organisations are composed of individuals and to conclude from this that all activists have to do, in order to achieve a change of direction in government and businesses, is to convince the majority of politicians, government officials and employees of businesses of the necessity for change.

This argument fails to recognise that organisations, whether they be government departments or business corporations, have structures and goals that make them behave in quite different ways from their individual members. For example, public companies are legally responsible to their shareholders, while private companies are legally responsible to their owners. In the Westminster system, government departments are responsible to their respective ministers, while in the US system secretaries and other senior managers of government departments are appointed by and responsible to the president. Even when the majority of members of a public or private company or a government department support strong climate action, the structures of these organisations and their goals generally preclude any change. The most prevalent organisational structure is that of the bureaucracy, a hierarchical system in which individual workers are interchangeable. This is an excellent system for carrying out routine operations efficiently under central direction, but it is very resistant to change from below.[4]

In a few rare cases, the most powerful person in a bureaucracy initiates change in the public interest, either as the result of external pressure such as a consumer boycott, or rational argument, or revelation. A famous example is the late Ray Anderson, founder and chairman of Interface, one of the world's largest floor covering companies, who experienced a revelation, a 'spear in the chest' as he described it, and made a 'mid-course correction' to take his company onto a sustainable development pathway.[5] This inspiring story is the exception that demonstrates the rule. Ray Anderson was by far the most powerful member of his organisation.

Another simple argument for relying on individual action is that, even if governments and big business do nothing, changes in many individual/household lifestyles can achieve the necessary social change. The argument is reinforced by a parable called *The Hundredth Monkey*. On an island there is a tribe of 100 monkeys. A single monkey learns a new technique, washing sweet potatoes, that makes a previously unpalatable food edible. The monkey teaches this new technique to another monkey, then both monkeys teach others and so on, until in theory all 100 monkeys in the colony are practising it.[6]

Human society (and probably monkey society too) is rather more complex than described in the story. Individual actions and innovations, provided they are spread from person to person, have some educational value within the human community, although the diffusion of knowledge in this way is generally very slow. As a result, individual actions have limited benefits. There is little point in exhorting people:

- to use public transport, if they live or work in an area that has inadequate public transport
- to avoid using electric resistance heaters, a big source of household greenhouse gas (GHG) emissions in regions where most electricity is fossil-fuelled, if the householders are tenants in a house that has no insulation or solar access or alternative low-carbon heating systems
- to purchase Green Power for an additional charge in cents per kilowatt-hour (c/kWh), when they are on low incomes
- to use solar power, if they live in rural areas at extremities of the electricity grid where the price of grid power is heavily subsidised both in terms of rate per unit of electricity purchased and the provision of free maintenance.

Furthermore, in many countries individuals and households are only directly responsible for a small fraction of energy use and GHG emissions. The major part generally comes directly from heavy industries and indirectly from the fossil energy embodied in the products we buy. This is one of the limitations of the idea of assigning a quota of direct GHG emissions to every individual and a debit card for withdrawing from this quota when fossil-fuelled electricity, gas or petrol is purchased. While such a system could make a useful contribution, educationally as well as in terms of cutting emissions, it falls far short of what is required.

Not only does the notion, that individual action is the principal transition pathway, fail to recognise the inability of individual action to change corporations, but it also ignores the lack of power of individuals to change laws, infrastructure and the economic system. Governments collect a large fraction of national wealth in the form of taxation and governments decide how to spend that wealth. They make the laws, which assist some actions and impede others. Will most transport funding go to roads or to public transport, cycleways and footpaths? Is a new power station really necessary and, if so, will it be fuelled by coal, gas or RE? Will all houses and commercial buildings be required to meet EE standards? Will our nation have a population policy? Will native forests continue to be logged? Will farmers be allowed to clear their land and, if not, will they be compensated? Where will a new business district or shopping mall of a city be located and will it be accessible by public transport? Will a proposal for a new aluminium smelter be approved and will its electricity price be subsidised, as is the usual practice?

In terms of GHG emissions, government decisions swamp those of individual people and individual households. Furthermore, large corporations influence all of the above decisions by government. Government in turn sets the framework in which corporations operate. Thus individual actions are constrained by much larger forces. This is even true on the domestic scale. If we live in rented housing, we cannot install insulation or a solar hot water system; that is the prerogative of the landlord.

That doesn't mean individual actions should be discouraged. Making our own homes more energy efficient, installing small-scale RE systems or buying Green Power, reducing our intake of beef and lamb, and driving less are all empowering and healthy. Therefore I have included an Appendix to the book that sets out in more detail what you as an individual can do to make your home and other activities more sustainable in terms of reducing GHG emissions. These actions set an example to friends and neighbours. To have credibility in asking others to act, we must demonstrate commitment to reducing GHG emissions in our own lives. Energetic and highly committed individuals can also play a vital role as champions of particular climate action campaigns and leaders of climate action groups. Individual actions are necessary and valuable, but not sufficient to meet the challenge of climate change. To gain effective climate change policies from government and business, pressure from a climate action movement is needed.

In the face of the formidable barriers to the Great Transition, we, as individuals, have two responsibilities for changing the energy system and reducing GHG emissions, additional to individual action.

The first and key responsibility is, as citizens, to exercise our democratic rights *collectively* to demand that our governments take rapid and effective action by implementing appropriate laws, regulations and standards, pricing and funding policies, education and information, industry policies, institutional change and population policies. As we shall see in the next chapter, our democratic responsibilities go far beyond voting as individuals once every three or four years. The collective actions of citizens cooperating are needed to change government policies. We can carry the greatest weight with decision-makers by joining and being active in citizens' groups, including environmental, RE, social justice, professional, trade union and faith groups.

Our second responsibility is to act as consumers to apply *collective* pressures to industry and business to supply ecologically sustainable and socially just products and services, in particular, sustainable energy. We can do this joining consumer and bulk buying groups and by encouraging these groups to take action to foster purchases of high-quality sustainable energy products.

A pathway to social change, that is likely to be more rapid and more influential than individual action, is direct action by community groups to implement cooperatively their own local energy production and efficient energy-use systems.

## Action by cooperatives and other community groups

Some organisational structures, notably cooperatives, are more responsive to the views of their members than companies. The concept of cooperatives builds on a European history of more than 150 years. Cooperatives are generally based on values such as self-help, self-responsibility, democracy, equality, equity and solidarity. Because of their member-driven nature, cooperatives support the social and economic development of communities. This means that decisions made by cooperatives balance the pursuit of profit with the needs and interests of members and their communities. Cooperatives take many forms and operate in all sectors of society: common kinds are financial cooperatives, agricultural cooperatives, trade cooperatives, housing cooperatives and energy cooperatives.

Energy cooperatives can open up medium-scale RE production by wind, solar, biomass and, in a few cases, geothermal sources. This is a scale of RE production that otherwise receives insufficient attention. They can have greater social and political impact than an equivalent number of individual household systems, by involving whole communities in ecologically sustainable energy generation and use. Thus they inform and educate local communities, stimulate media interest, create local jobs and involve local politicians and local governments.

Energy cooperatives have played important roles in the development and dissemination of RE in several countries, notably Denmark and Germany. Denmark led the revival of wind power in the 20th century. The oil crisis in the 1970s and public opposition to nuclear power motivated the publication of two energy plans by independent groups of experts. By the 1980s several local manufacturers were selling wind turbines with capacities of 55 kW upwards. Since these were too expensive for most individuals, local wind cooperatives were developed, assisted initially by subsidies on the capital costs of wind turbines and the legal requirement for utilities to pay a fair price for wind energy fed into the grid. By

1996, there were around 2100 cooperatives throughout the Danish countryside, which created the basis for continuing popular support for wind power. By 2001, wind turbine cooperatives, including more than 100 000 families, had installed 86 per cent of all turbines in Denmark. In addition, local distribution of electricity has become the responsibility of local not-for-profit cooperatives, local government, or companies with a concession.[7]

In Germany over 300 new energy cooperatives with more than 80 000 members have been established since the mid-2000s. The liberalisation of the energy market and the introduction of feed-in tariffs through the Erneuerbare-Energien-Gesetz (EEG – Renewable Energy Act) provided incentives for private investors, both individuals and groups, to engage in electricity production. Most of the cooperatives formed in response to this Act are in small villages and towns. According to the utilities Deutscher Genossenschafts- und Raiffeisenverband e.V., with total annual electricity production of 290 gigawatt-hours (GWh), the production capacities of energy cooperatives exceed the electricity demand of their members.[8]

Energy cooperatives generally can be distinguished from one another by the way they consume, produce or trade energy in the form of electricity or heat. For example, in Germany initiatives have been developed to finance solar photovoltaic (PV) systems on the roofs of public buildings, such as schools, fire stations, community halls and churches. In these cases the cooperative structure enables different individual participants with limited investment capacities to get together and make use of vacant public spaces that a single person/household could not access. The same applies to farmers lacking financial resources but able to provide rooftop spaces on sheds and other storehouses for PV systems owned by groups. Energy cooperatives have evolved in big cities in order to make use of public spaces for solar PV systems (eg, Potsdamer Solarverein e.V.).[9] Other community-driven activities in Europe are cooperatives that establish district heating systems, energy self-sufficiency villages and towns based on bioenergy (such as Güssing in Austria)[10] and community-owned wind farms. Some communities generate so much RE that they can provide for their own energy needs and export a surplus – an example is the island of Samsö in Denmark, which generates electricity from the wind and provides district heating by burning straw residues from its agriculture.[11]

Cooperatives in Germany are also beginning to invest in local electricity distribution and long-distance power transmission. The cooperative BürgerEnergie Berlin is raising funds to bid for the purchase of the local distribution system in Berlin when the contract of the existing owner, Vattenfall, expires.[12] Recently the German Environment Minister announced plans to allow citizens to invest in the construction of transmission lines to serve new RE farms as old nuclear power stations are shut down. Since part of the profit, that giant power utilities would have made, will go to consumers, the minister is hoping that the scheme will reduce opposition to the new power-line constructions.[13]

In the USA, distribution cooperatives, which deliver electricity to retail customers, are the foundation of the rural electricity network. There are also generation and transmission cooperatives, which provide wholesale power to the distribution cooperatives, either through their own generation or by purchasing power from other suppliers. As of March 2012 the National Rural Electric Cooperative Association (NRECA) had 905 member cooperatives serving an estimated 42 million people in 47 states. They generate 5 per cent of total annual electricity produced each year, deliver 11 per cent of the electricity and employ 70 000 people in the USA. Unfortunately NRECA's website shows little interest in taking action on climate change or promoting RE.[14] However, the National Renewables Cooperative Organization was founded in 2008 to promote and facilitate the development of RE resources for members. It has 25 member cooperatives and so far a few of these members are focusing on purchasing power from wind farms.[15] The huge potential for RE cooperatives in the USA still remains to be harvested.

## Action by corporations

Big business with vested interests in GHG pollution has been, and still is, the major source of funding for groups that deny climate science, attack climate scientists and the Intergovernmental Panel on Climate Change, and oppose the transition to sustainable energy. To this end big business lobbies politicians and public officials, runs media and advertising campaigns, and produces and disseminates myths to undermine climate science, EE and RE.[16]

One of many lobby groups against climate action was the Global Climate Coalition, which influenced the US Senate to pass a unanimous resolution in 1997 that the USA should not be a signatory to any protocol that did not include binding targets and timetables for developing as well as industrialised nations or 'would result in serious harm to the economy of the United States'. In 2002, after President George W Bush announced that he would reject the Kyoto Protocol, the Coalition disbanded, having achieved its specific goal.[17]

Nowadays, campaigns of RE denial have supplemented climate science denial. A network of right-wing organisations, including Americans for Prosperity, which is funded by the billionaire Koch brothers (whose industrial interests include energy and minerals), is attacking President Obama for his support for RE. The American Legislative Exchange Council, which also has financial links to the Kochs, has drafted bills to overturn state laws promoting wind energy. *The Guardian* reported on a confidential strategy memo that advises using 'subversion' to build a national movement of wind farm protestors. *The Guardian* also reported that the strategy proposal was prepared by a fellow of the American Tradition Institute (ATI) – although the thinktank has formally disavowed the project.[18]

Despite such well-funded campaigns by vested interests, a few large corporations with long-term visions have at various times made credible public commitments to sustainable energy. From the 1980s, BP Solar, a subsidiary of the oil and gas giant BP, manufactured solar PV modules in several countries. From about 2000 the parent company branded itself as 'Beyond Petroleum', withdrew from the Global Climate Coalition, installed PV on the roofs of its petrol stations, and initially its profits soared. However, following a series of serious environmental and safety violations and disasters, culminating in the 2010 explosion and oil spill at the Deepwater Horizon drilling rig in the Gulf of Mexico, it lost public credibility. In 2011 BP closed down BP Solar, saying that it would in future concentrate on wind and biofuels. Even before the closure RE was a tiny part of BP's operations.

Supermarket giant Walmart had the goal of reducing GHG emissions from its stores, clubs and distribution centres worldwide by 20 per cent by 2012, compared with its 2005 baseline level. To do this it planned to implement EE, improve refrigeration technology and implement RE. By the end of 2010, it had achieved a reported reduction of 12.74 per cent, but has not yet published subsequent data. It is also working with its suppliers to assist them to reduce their emissions.[19]

Google claims that, as of 2011, it has made its data centres twice as energy efficient as the average, that one-third of the electricity used by its data centres is provided by renewable sources and that GHG emissions from the remaining two-thirds are neutralised by offsets. It has committed over $1 billion to RE projects that create far more RE than it consumes.[20]

Unfortunately the Google case study and even the more modest efforts of BP and Walmart in support of sustainable energy are not typical of the corporate sector, even though there are substantial economic savings to be made from improving EE in large, energy-consuming corporations. These savings would be even greater if large electricity consumers did not receive special low electricity rates in some countries, in effect, cross-subsidies from household electricity consumers.[21] In addition, there would be a much wider uptake of solar PV by corporations with high electricity consumption and extensive roof areas (on supermarkets, warehouses, data centres and car parks) if governments cut a fast pathway through the red-tape that inhibits medium-scale RE project proposals in many countries.

In the long run, the ongoing reduction in the costs of RE systems, increasing costs of coal, oil and electricity from the grid, and the growing perception of the financial risks of investment in fossil fuels will encourage business and industries other than the big GHG polluters to invest more and more in sustainable energy. The insurance and reinsurance industry, already feeling the impacts of climate change, is pushing for climate action. Several RE manufacturers are now quite large corporations, each employing thousands of people. But relying on markets is a slow process, especially when huge economic subsidies to the production and use of fossil fuels are still available (see Ch 9, 'Removing subsidies to fossil fuels and nuclear energy'). Corporations will move rapidly if governments structure the business environment to allow business to make a profit from investing in sustainable energy and a loss from failing to do so.

## Conclusion

Individual action – to improve EE in the home, install small-scale RE systems or pay for Green Power, and reduce car use – is necessary. Everyone can do something of value. But, individual action is often constrained by lack of home ownership, inadequate public transport and low incomes. Even if these constraints could be removed, individual energy use is a fraction of total energy use and so individual action is not sufficient.

In many countries substantial potential exists for cooperatives and other local community groups to install medium-scale RE systems. These projects could greatly expand the market for RE while building community engagement and support. They can foster the social and economic development of localised communities, stimulate media interest and involve local government politicians. While energy cooperatives are very successful in Denmark and Germany, in many other countries and states they are constrained by lack of facilitating legislation, financial mechanisms and other institutions.

Large-scale RE production and use is highly dependent on policies of national and state governments. Policies are needed to create a business environment that encourages a transition to sustainable energy. Leaving it to the market will not realise the benefits in time to avoid serious and probably irreversible global climate change, local air pollution and energy insecurity.

With strong and appropriate government policies, as outlined in Chapters 8 and 9, individual, community and business action to implement sustainable energy would become much easier and faster. The key remaining question, addressed in the next chapter, is: how can citizens apply pressure to governments to implement effective measures to rapidly expand sustainable energy?

# 11: CITIZEN ACTION

Change comes from power, and power comes from
organisation. In order to act, people must get together ...
Power and organisation are one and the same.

Saul Alinsky[1]

It is night. I'm a passenger on a crowded boat hurtling down a
wild river. In the distance, we hear a roaring sound. Worried,
some of the passengers turn to the captain and crew. 'That sounds
like a waterfall ahead! Shouldn't we turn to the shore and take an
overland route?', we ask. 'Sit-down and relax', the captain replies.
'My map shows that this is the fastest route to El Dorado. And it
doesn't show any waterfalls on this river.' Other passengers call
out: 'The captain is in charge and his map was drawn by experts:
economists and businesspeople.' The roaring sound grows louder,
nearer. Then the boat enters a dense cloud of spray. Now thor-
oughly alarmed, the concerned group of passengers try to seize the
tiller, but are forced back by the crew who are armed. Terrified,
I wake up. Thank heavens it was just a dream!

Chapters 8 and 9 have discussed the key policies we need from
political leaders to protect human civilisation and the millions
of other species that inhabit Earth from the ravages of human-
induced climate change. Just a few countries and states/provinces
are taking substantial, effective action, notably Germany, Denmark,
California and, until recently, Spain. A growing number of towns
and cities have set targets of very low to zero greenhouse gas
(GHG) emissions. China, now the world's biggest GHG emitter,

is simultaneously following the two mutually contradictory paths of climate destruction by fossil fuels and climate action by renewable energy (RE). On a national scale the USA, the second biggest emitter, is paralysed by a divided Congress. At this critical stage there is negligible progress in international meetings and the vast majority of governments are implementing actions that are token at best – and regressive at worst. They tend to identify with the huge industries that are destroying the planet, rather then the new, cleaner, small to medium-sized industries and community groups that are trying to save it.

Among the recalcitrants special mention must be given to the federal and state governments of my own country, Australia, the world's biggest coal exporter and the country with the highest per capita GHG emissions among developed countries. Australian governments, federal and state, are supporting the expansion of coal mines, port facilities and freight railways, aspiring to triple exports of coal and to greatly expand exports of liquefied natural gas and coal seam methane. They pay lip service to climate action, but do not allow that to influence policies to further expand fossil fuels and minerals production. The Coalition Party, which has become the government in September 2013, is even trying to terminate the modest carbon price implemented by its predecessor, to close the Clean Energy Finance Corporation and to reduce the Renewable Energy Target. At the behest of a noisy minority who are spreading fear of a fictional 'wind turbine syndrome', several state governments are putting restrictions on wind farms, so severe that they have almost stopped the growth of this essential, environmentally sound industry. The Murdoch press, which dominates the Australian press/media, is full of articles denying climate science and attacking RE. The majority of the population, confused by propaganda from the deniers, is still sympathetic to RE, but unsure as to whether it can replace fossil fuels and so is reluctant to speak out.

In the face of public confusion, ignorance and complacency, and the corruption of democratically elected governments by the enormous wealth of the industries with vested interests in GHG pollution, what are concerned citizens to do? The climate action movement has a well-reasoned case, based on climate science and the ethical position that we, the present generation, should not leave a degraded planet to our children and grandchildren. However, the failure of governments to establish and implement strong international agreements and to act effectively at the national and state/provincial levels, demonstrates that science and ethics are not sufficient to overturn the existing system. We must recognise that the struggle for climate mitigation and a sustainable energy system is fundamentally a power struggle, one that will determine the future of human civilisation.

## The unavoidable power struggle

Although it is facing two formidable foes – wealthy industries with vested interests and a complacent public – the climate action movement has three principal strategic strengths:

- It can potentially call upon the support of far greater numbers of people than vested interests in greenhouse pollution, the self-styled Greenhouse Mafia, to carry out its campaigns.
- It serves, with integrity, humankind as a whole, instead of the owners and shareholders of destructive industries. This provides it with a moral advantage and credibility that flows from it.
- It is sufficiently diverse to influence groups and individuals in almost all walks of life.

By building upon these strengths, the climate action movement can increase its power to match that of the Greenhouse Mafia. Social change organiser Saul Alinsky writes that the following goal must guide the strategies and tactics of the movement:[2]

> ... every move revolves around one central point: how many recruits will this [action] bring into the organisation, whether

by means of local organisations, churches, services groups, labor unions, corner gangs, or as individuals. The only issue is, how will this increase the strength of the organisation ... Change comes from power, and power comes from organisation.

In order to act, people must get together ... Power and organisation are one and the same.

So the key challenge for the climate action movement is to recruit and organise large numbers of concerned citizens and build strategies to enable them to challenge the power of the Greenhouse Mafia.

The opening sentence of Gene Sharp's classic book, *The Politics of Nonviolent Action*, reinforces this: 'Some conflicts do not yield to compromise and can be resolved only through struggle.'[3] He goes on to write (p 7):

Unlike utopians, advocates of nonviolent action do not seek to 'control' power by rejecting it or abolishing it. Instead, they recognise that power is inherent in practically all social and political relationships ... They also see that it is necessary to wield power in order to control the power of threatening political groups or regimes.

Sharp's view is based on a detailed study of nonviolent actions over millennia from around the world. He states (p 8) that, although governments appear to exercise 'monolithic' power and that the people appear to be 'dependent upon the good will, decisions and the support of the government or any other hierarchical system to which they belong', reality is the opposite. Drawing upon many historical examples, Sharp argues that governments actually 'depend on people, that power is pluralistic, and that political power is fragile because it depends on many groups for reinforcement of its power sources'. The power of government and corporations depends upon the consent of the people – and we the people can together withdraw that consent. In this context, Sharp is referring to a much wider range of actions than voting at elections.

Box 11.1 explains that, if the climate action movement is to be successful, its tactics must be nonviolent. A large and growing body of experience exists in nonviolent social change movements in fields of social justice, environmental protection, consumer rights, community development and peace. Some of the leading groups and individuals have been recognised by the Right Livelihood Awards,[4] which are sometimes described as 'the alternative Nobel Prize'. The accounts of these successes are inspiring and I recommend them as an effective tonic when the power of vested interest seems overwhelming.

Some readers may feel uncomfortable with this talk about power and the need to target it against the individual decision-makers. 'Surely', you may say, 'all we have to do is give the decision-makers the facts and they will make the right decisions.' Unfortunately we have seen that reason and sound ethics are not sufficient to sway the political power-holders, who are in thrall to vested interests. 'Even if I accept that', you may add, 'in a democratic system we can vote for the party with effective climate action policies.'

The response of experienced climate activists is that in several nominally democratic countries the choice at election time is between a party opposed to action and a party supporting weak ineffective token actions. Even getting weak legislation through can be very difficult. The parties with strong climate action policies are usually minorities, such as the Greens. Furthermore, the democratic process is not limited to voting once every 3–5 years. The Greenhouse Mafia are campaigning 365 days per year, every year. To counter this, the climate action movement must spend the time between elections recruiting and organising to build its power base, developing strategies, engaging with the media and carrying out a wide range of tactical actions to resist the greenhouse-polluting industries, build clean and safe alternatives and further strengthen the movement.

In the 'Further reading' section you will find a number of excellent books on the nuts and bolts of building social change movements. In the following sections of this book I draw upon these sources to outline the key issues: organising, strategy and tactics.

## 'Organising' groups

In this chapter, the word 'organising' conveys both the standard meaning – to create a structure for a group – and a special meaning used by some activist individuals and groups for the activities of an organiser (described below), which include developing strategy and planning a campaign. Both kinds of organising are important for climate action groups (CAGs), so that they can build membership, spirit, trust, resources and capacities of their members and the group as a whole to develop and carry out a campaign plan. CAGs are working for a better society and so their own structure and operations should be designed to achieve this.

Most successful CAGs do not grow and take effective actions without at least one guiding spirit: the organiser. In the context of social change movements, an organiser is a person who facilitates a community to empower itself. He or she does this by:

- guiding the formation and growth of one or more CAGs
- helping the group to develop a shared vision, strategy and tactics
- fostering a democratic group structure and decision-making processes
- nurturing leaders and
- guiding the organisation's public meetings, workshops, study groups and actions.

Thus the organiser is somewhat different from a leader, whose role is to lead people into action. The organiser brings individuals together, helping them to build organisations and campaigns. This person facilitates from behind the scenes, while the leader is up front in the public eye.

Organisers of many small community groups work in a voluntary capacity. However, the paid (usually under-paid) organiser has a well-recognised role in the trade union movement and the concept has been broadened to the wider social justice movement by the organiser and activist Saul Alinsky,[5] among others. Nowadays the organiser plays an increasingly important role in the environmental movement too. Incidentally, before Barack Obama became a politician, he was a community organiser in a poor black area of Chicago. Unfortunately, as president he seems powerless to implement the ideals he lived out as a young organiser and expressed in his first autobiography.[6]

## Strategy

Strategy is the planning and conduct of long-term campaigns to achieve broad goals. Nonviolence expert Gene Sharp describes strategy as 'charting the course of action which makes it most likely to get from the present to a desired situation in the future'.[7] In its book *Organizing for Social Change*, the Midwest Academy, a training institute for progressive social change, defines strategy much more specifically as 'an approach to make a government or corporate official do something in the public interest that he or she does not otherwise wish to do'.[8] This definition deliberately targets one or more individual decision-makers rather than an organisation. It states that strategy requires your organisation to exercise power against that individual. As I see it, this offers a valuable perspective, but is not sufficiently broad for all the needs of social change movements. In several countries the policies of a political party or a business corporation are rarely determined by a single person.

One of the first strategic issues to be discussed by a new CAG is the formation of its goals and the issues it will campaign on. Most CAGs will choose several issues, some of which can be achieved within a few months, in order to post some wins on the board, and one or two more difficult, long-term issues, to stretch the group. Issues should be evaluated carefully in a participatory process. According to activist groups, they should satisfy most of the following criteria:[9]

1   Improves people's lives.
2   Builds people's confidence in their own power.
3   Strengthens the power of the people in relation to the decision-makers.
4   Is winnable.
5   Is widely and deeply supported in the community.
6   Can be presented clearly and simply to the media and public.
7   Has one or more particular individual decision-makers as a clear target.

It may be helpful to apply a SWOT analysis to proposed goals and issues. This is a strategic planning method to identify the Strengths, Weaknesses, Opportunities and Threats/Barriers for the goal. Strengths and Weaknesses involve the group's internal resources in relation to the goal. Opportunities and Threats involve conditions external to the group that may influence its ability to reach its goal. The results of the SWOT analysis may indicate that one of more of a group's goals is unattainable. Thinking positively, SWOT analysis should lead the group to ask:

• How can we use each Strength?
• How can we address each Weakness?
• How can we seize each Opportunity?
• How can we avoid, diminish or counter each Threat?

Strategy may also benefit by reflection and discussion about the psychology of fostering social change. Although we must inform people about the impacts of climate change on human society and other species, we must understand that many people are not motivated by placing heavy emphasis on the threats and potential disasters from business-as-usual. This can simply paralyse people, fostering denial instead of constructive action. The idea of a radical change in lifestyle can also be scary, making them feel insecure. So, what is the solution? One approach is that of social marketer Les Robinson, who has designed a theory of successful social change with six ingredients:[10]

- Positive buzz – when people share optimistic stories about change.
- An offer of hope – when people make the connection between a novel action and their own hopes and frustrations.
- An enabling environment – when people's environments make new behaviours easy to do and sustain.
- A 'sticky' solution – when behaviours are reinvented to better fit people's lives.
- Expanded comfort zones – when people are helped to reduce their fears.
- The right inviter – when inspiring, trusted peers invite action.

In developing a social change strategy, climate activists could also draw lessons from the social psychological and sociological approach that considers the way people are influenced by messages from the mass media, advertising, films and peers (see Ch 7, 'Social psychology of travel behaviour').

## Tactics

President Franklin D Roosevelt once said to a group of activists, 'Okay, you've convinced me. Now go out there and bring pressure on me.' To achieve social change, clearly we have to mobilise the support of the majority of the population. But we also need to ensure that the power-holders understand the issues and necessary policies. If we lobby without grass-roots support, this will carry little weight with power-holders. On the other hand a mass movement will not gain the right decisions from the power-holders unless it has communicated its requests to them, usually by lobbying. For this reason, the climate action movement needs a wide range of tactics from different CAGs. Some tactics can be directed towards influencing the power-holders directly, others towards the community at large and various key subgroups within it, and still other tactics towards important intermediaries such as the media.

Tactics are the individual steps or tools used in carrying out a strategy. Tactics are limited, short-term courses of action on the long-term strategic pathway. They may involve face-to-face lobbying, media events, an educational campaign, a lawsuit and sit-ins, for example.

Most tactical actions will be 'soft', in the non-confrontational sense. They are designed to win people around to the value of and need for the proposed changes. However, not all power-holders can be won over by soft words, especially if they identify strongly with the vested interests that maintain the status quo. There are still some politicians and business leaders in key positions who are actively opposing climate action. So, some actions will have to be confrontational. These nonviolent confrontations are not intended to win over diehard power-holders, but rather to resist the latter's bad decisions, educate the public and create media events.

Before discussing individual tactics, the CAG must consider how to *frame* the issue or campaign, that is, how to present it to the public in a context and language that you choose.

### 'Framing' the campaign

Communication is a vital part of any campaign. The CAG must decide how it wishes itself and its campaign to be seen by others, including decision-makers, influential organisations such as business, trade unions and professionals, other non-government organisations (NGOs), the media and the community at large. To some extent it can influence its 'image' by the way it describes itself and its vision, its goals, its campaign and the issues as it sees them. Its website, press releases, media interviews, published articles, social media engagement and communications with individuals all help to do this. This process is known as 'framing' and it involves choices of conceptual framework for the campaign and the language used.

Opponents, the media and other interested parties each have their own agendas and each will try to place the CAG and its issues into a 'frame' of their own making. It is common to label individuals and groups who are pushing for constructive social change as 'emotional', 'ideologically driven', 'irresponsible', 'radical' and even 'unscientific'. Of course, much of this is projection in the psychological sense. There is no-one more emotional than the CEO of a polluting industry who is facing the higher costs of environmental protection and the resulting reduced profits. There is no-one more ideologically driven than a narrow neoliberal economist who tries to impose the competitive market model on parts of the economy that are manifestly suffering from market failure. No-one is more irresponsible or radical than a business leader or politician who wishes to risk the survival of human civilisation for short-term profits or the flawed ideology of endless economic growth on a finite planet. No-one is more unscientific than a climate change denier who claims, contrary to all the empirical evidence, that the majority of glaciers are actually expanding instead of shrinking.[11]

Linguist George Lakoff has emphasised the way language shapes how people think.[12] Drawing upon his advice climate activists learn to reframe the terminology used to market environmentally and socially damaging processes: for instance, the pejorative term 'command and control' used frequently by economists can be reframed as the more neutral 'regulations and standards'; 'burden sharing' used by those opposed to carbon pricing can be replaced with 'the distribution of costs'; 'clean coal', a marketing term for carbon capture and storage (CCS), an unproven technological system for coal power, can be replaced by 'CCS'. Even the term 'global warming' has cosy overtones to someone who is unaware of the current impacts and the potential future disasters. In view of the melting of ice-caps and glaciers, floods, droughts, storms, firestorms, etc, 'climate disruption' or 'climate crisis' or even 'climate emergency' are more appropriate terms.

### Campaign tactics

A CAG chooses campaign tactics from the wide range of options available (Table 11.1) on the basis of its size and resources, the

TABLE 11.1    **Tactics available to CAGs**

| Actions | Actors |
| --- | --- |
| Lobbying power-holders | Large generalist environmental non-government organisations (ENGOs) |
| Building alliances and partnerships with influential organisations | Large generalist ENGOs |
| Research | Many groups research the issues, often assisted by academic members or advisers |
| Education and information for particular groups and members of the public | All |
| Networking among ENGOs | Medium–large ENGOs and peak organisations of ENGOs |
| Community energy projects (see Ch 10, 'Action by cooperatives and other community groups') | Dedicated community groups, local governments and small and medium-sized business enterprises (SMEs) |
| Publicity via the media and social media | All. Local, state and national groups generally obtain media on local, state and national levels, respectively |
| Legal action (impossible in some countries) | Large ENGOs and alliances |
| Nonviolent confrontational action | A few large generalist ENGOs, such as Greenpeace, and specialist small CAGs |

interests of its members, and the policies and attitudes of the governments in its jurisdiction. CAGs may be formed by environmental, social justice, professional, trade union, business, political party, faith or other community organisations, or may be local groups with a diverse membership.

Now let's consider some specific tactics.

## Lobbying decision-makers

Lobbying is defined to be communication with decision-makers who are addressed specifically by name or position. In this sense, lobbying covers face-to-face meetings, phone calls, emails, letters, petitions, submissions to government enquiries, addresses to public hearings and accountability meetings. Lobbying is a legitimate and important part of participatory democracy. Since vested interests place considerable resources into lobbying,[13] some CAGs need to participate as well.

Decision-makers to be lobbied include:

- ministers, shadow ministers and their advisers
- other members of parliament/congress in all spheres of government and their advisers
- senior public officials
- chairs and members of parliamentary/congressional committees
- party committees and
- CEOs and other senior managers of businesses, peak organisations and other bodies.

Keep in mind that, to be effective, lobbying must be done from a position of some strength. Politicians and political advisers should be made aware of the number of potential votes the delegation represents. CEOs of corporations should be made aware of the ability of the people represented by the delegation to decrease or increase their sales through consumer action and/or media publicity. Public officials may respond positively to lobbying if they fear a media campaign exposing that their decision-making on the issue did not follow due process.

### Nonviolent confrontational action

> Nonviolence is a powerful and just weapon ... which cuts without wounding and ennobles the man who wields it. It is a sword that heals.
>
> Martin Luther King, Jr[14]

In the context of climate action, nonviolent actions are undertaken to confront key decision-makers in polluting companies and developments, the financial institutions that fund them, and the governments that are complicit with them. These confrontations are used to resist a bad government policy or a harmful project, such as a new dirty coal-fired power station or coal mine; to mobilise community opposition; to demonstrate to government the growing community concerns about the issue; or all of the above. Media publicity is a valuable aspect, but is rarely the principal objective of the confrontation, which is to exert pressure on the decision-maker as part of the process of winning a demand.

---

### Box 11.1  Why protest actions must be nonviolent

Social change activists need to avoid using physical violence for the following reasons:

- Most people will not participate in violent actions.
- Violence distracts media attention away from the issue to the violence.
- Violence alienates the wider community and isolates the campaign.
- The state, backed by the military and the police, has by far the greatest capacity to conduct violence. Therefore, any violent action that falls short of a well-resourced popular revolution will inevitably fail.
- Human-induced climate change, together with many other issues of environmental protection and social justice, are the result of violence towards nature and people that is implicit in the actions of power-holders. Meeting violence with violence is morally untenable for solving such problems.
- In practical terms too, violence is untenable because it simply fosters more violence in an endless cycle that is extremely difficult to break. In a quotation attributed to Mahatma Gandhi, 'An eye for eye only ends up making the whole world blind.'[15]

Power-holders are well aware that the use of violence weakens a social movement both internally and in the eyes of the wider community. Therefore they sometimes send agents provocateurs to foment violence that is blamed upon the protest group.[16] For

---

this reason, training in nonviolent action is important for CAGs that are planning nonviolent actions. It can even include methods of nonviolently isolating, surrounding and limiting the actions of a violent minority in a crowd.

Nonviolent action training has a wide range of other benefits, as discussed in *Resource Manual for a Living Revolution*.[17] For example:

- It introduces cooperative ways in which people can learn about and change their world.
- It develops skills in conflict resolution and democratic decision-making.
- It allows skills, ideas and organising methods to be developed and tested in practice situations where risks are low and mistakes are less costly.
- It teaches methods of creating trust and solidarity than can be effectively applied to withstand discouragement and repression.

Nonviolent actions include rallies, marches, sit-ins, pickets, naming and shaming, divestment, shareholder actions, strikes and boycotts. Rallies, marches, sit-ins and pickets are ways campaigners can physically, but nonviolently, block or occupy a site, such as a road, public space, gateway, office or whole building.[18] The larger the number of protesters, the longer the event can be maintained in the face of police opposition and the more it will attract media attention. Increased numbers will also apply more pressure on targeted decision-makers.

Individuals and organisations that are climate reprobates can be named and shamed in various ways. For example, the British Royal Society published its letter to Exxon Mobil, objecting to the giant oil company's funding of groups that were denying greenhouse science and undermining climate action by governments.[19] Another effective method of naming and shaming is to award a prize for (say) the worst greenhouse-polluting company, or the politician who has given the most support to greenhouse pollution, or the land-owner who has cleared the most land of native forest. An amusing well-staged event can expose a serious problem and its perpetrator, while gaining brief but positive media coverage for the campaign. For all naming and shaming, it is essential to be sure of the facts and to avoid defamation, for example by casting doubt on someone's motivations.[20] It is entirely proper to point out that a Minister for Energy holds shares in a coal company and is therefore in a position of potential conflict of interest. However, it could be defamatory to suggest that the minister's approval of a new coal mine was influenced by his/her shareholdings. If there is any doubt, first seek legal advice before making a public statement targeting an individual or corporation.

Divestment (withdrawing investments from environmentally or socially damaging industries) is currently becoming a major part of the international climate action campaign. It is challenging educational and religious organisations, city and state governments, and other organisations that purport to serve the public good to divest from fossil fuels.[21] By mid-June 2013 in the US ten big cities have announced plans to divest and there are active struggles for divestment on 380 college campuses. Bill McKibben, the writer-activist founder of 350.org, points out that divestment is a moral issue – 'If it's wrong to wreck the climate it's wrong to profit from that' – as well as a financial issue:

> A decision to invest in fossil fuel shares at this point is a bet
> that the planet will do nothing at all about climate change. If
> the world's governments ever took even that small two degree
> target seriously, as HSBC and Citi noted in a report last month,
> the share values of these companies would be cut in half.[22]

Some environmental NGOs and socially responsible financial institutions are buying a few shares in greenhouse-polluting companies. Then they can attend shareholders' meetings and ask awkward questions of the chairman of the board. This is a way of challenging the board and other shareholders to be more socially responsible.

Another possible tactic, proposed by Saul Alinsky,[23] is to ask shareholders in a company to assign their proxy votes to a CAG. An important part of the process is to lobby and negotiate for proxies with large institutional shareholders in the company. These could be other companies, superannuation funds and in some cases universities. In addition, it may be possible to acquire a large number of individual proxies through internet NGOs such as Avaaz and GetUp!, churches, clubs and societies, and environmental NGOs. If enough proxies can be obtained, then the opportunity arises to actually put motions and influence votes at shareholders' meetings.

To discourage a financial institution from lending money to a project that would be high in GHG emissions, a CAG could organise a campaign to encourage depositors to withdraw their savings from the financial institution on a specified date, if it does not withdraw from the proposed loan. Like acquiring proxies, the intention of this tactic is to gain a stronger negotiating position for the CAG. The challenge for the CAG is to use all means of publicity – websites, social media, email lists and traditional media – to mobilise people for the potential withdrawal. Banks only hold typically 10 per cent of deposits in reserve; the remaining 90 per cent is lent to investors. Therefore, a large total withdrawal on a particular date would reduce the bank's ability to make loans and hence its profit margin. A very large total withdrawal could force it to borrow or sell assets to avoid running out of its reserves and defaulting on its obligations. As a bonus, the publicity from this tactic is a valuable part of a public education campaign.

Before considering legal action a CAG needs two things: suitable legislation and competent lawyers willing to undertake a case in the public interest either pro bono or at low cost. Given these conditions, court cases may be a useful tool in some jurisdictions. Nevertheless, it must be recognised that successful legal actions by community groups against governments and large corporations are rare. All the advantages are with those who have money and power.

# What we can do

Dear Mark,

This morning we heard you speaking on the radio about your university's research on how we could run the Australian electricity system entirely on renewable energy. That is terrific news and it gives us hope! We are very concerned about climate change and would like our children to be able to live in a stable climate with safe, non-polluting, renewable energy. How can we assist the transition? We want to do something, but are not technically skilled. We run a small business and live in a rented home.

Best regards,
Vandana and Mike.

Dear Vandana and Mike,

Thank you for your email. As I see it, the barriers to a safe climate and a sustainable energy system are no longer primarily scientific or technological. Instead, they are political and cultural. We already have all the technologies we need, although there is still need for research and development to foster continual improvement. The key priority is to build an irresistible grass-roots movement to push for the transition.

Groups of concerned citizens can achieve much more than their members could as individuals. Therefore, I recommend, first and foremost, that you join or support a group that is informing the community about the threat of climate change and the opportunities offered by energy efficiency and renewable energy. Even better, you could consider joining a group that is going beyond education and information to apply pressure to decision-makers in government and big business to take effective action to drive the necessary transition to sustainable energy. You may have the choice of national, state and local groups that are part of an international movement, such as 350.org or a large environmental NGO or a local branch of the International Solar Energy Society, or a dedicated local climate action group.[24]

Joining a group that's committed to the public good can be very rewarding in human terms as well as progressing the movement, because you can meet and cooperate with some wonderful people.

If you are members of a political party that's half-hearted about climate action, you could try to influence policy from within. If the party is actually hostile to climate action, you could resign and join one of the few political parties that is strongly committed.

Although your home is rented, you can still use a wide range of technologies to save energy and so reduce emissions. To name just a few: compact fluorescent or LED lightbulbs; water-efficient showerheads; insulating curtains; draught excluders around windows and doorways; an insulating blanket for an indoor hot water tank; microwave cooker; solar clothes drier (aka clothesline); energy-efficient appliances; and a laptop computer instead of desktop.[25]

Behavioural change is important too. If your electricity supplier has a Green Power scheme, pay a little extra to get 100 per cent renewable electricity. Join a cooperative community renewable energy project if there is one in your region. If your home has a reverse-cycle air conditioner, operate it less frequently and wear warm clothes in winter and use fans in summer. In winter, heat only the rooms you are occupying. When taking short trips, leave your car at home and walk or cycle – you will be healthier too. Wash your clothes (except reusable nappies/diapers) in cold water. If possible, shift your savings and superannuation to a financial institution that does not invest in fossil fuels.

Many of the above technologies and behavioural measures will save you money as well as energy. Show your relatives and friends what you are doing. Spread the word. And do keep me informed of your progress.

Best wishes,

Mark

# 12: CONCLUSION

> You cannot save the world through incrementalism. There are
> lots of incremental things we can and should do to make the
> world tick a little bit better, but they're not going to save it.
> We need radical transformation.
>
> Peter Bakker[1]

Human civilisation is facing two great threats: global warming, caused primarily by the emission of greenhouse gases (GHGs) from the combustion of fossil fuels, and nuclear war, which could devastate our cities, industries and people with blast, firestorm, irradiation and then, by screening the Sun with vast quantities of soot, initiate a nuclear winter that could decimate agriculture for several years. Both threats could kill billions of people; global warming gradually and nuclear war quickly.

We now have technologies and measures that can solve a major part of the global warming threat. Most of them are already commercially available, although the scale of their manufacture and use must be multiplied many times over and quickly. Those that can do the job without increasing the risk of nuclear war are energy efficiency (delivering the same energy services with less energy), energy conservation (reducing the energy services we demand) and renewable energy (RE). They are compatible with an ecologically sustainable and socially just society. Their costs have been declining rapidly as their markets expand. Global annual investment in renewable electricity, including large-scale hydro, amounted to $260 billion in 2012, approximately equal to investment in fossil fuel generating capacity ($262 billion). However, net investment in fossil power, which does not include the replacement of retired power stations, was only 57 per cent of renewable capacity.[2]

These safe and everlasting technologies and measures can also bring substantial benefits additional to stopping global warming. They can reduce the pollution of air and water and associated health hazards, reduce land degradation, protect biodiversity, establish long-term energy security and foster greater national and local energy autonomy.

Several countries – notably Denmark, Germany and Scotland – several states, and many towns have set targets of 80–100 per cent RE and are working to achieve their targets. But the majority of countries are still making small, token gestures towards cutting GHG emissions and making the transition from fossil fuels to sustainable energy. Why the delay? The science is clear: if we are serious about limiting global warming to 2°C (which is not a safe level), we must leave most of the known reserves of fossil fuels in the ground. The ship of human civilisation is heading down a mighty, fast-flowing river towards a waterfall and, if we don't turn it around by about 2020, it will almost certainly go over the edge.

The problem is twofold. The majority of the passengers are either unaware of the gravity and urgency of the climate crisis or in a state of denial and hence are unwilling to give it priority in their concerns. The captain and officers of our ship are dominated by wealthy, powerful vested interests, the self-styled Greenhouse Mafia. Not only do they deny climate science, but they are also spreading misinformation about sustainable energy. This book busts the myths and exaggerations they peddle, compiling the evidence that:

- Efficient energy use, together with energy conservation, is the fastest and cheapest way of dramatically cutting GHG emissions from the energy sector.
- The so-called 'rebound effect', in which money saved through saving energy is allegedly spent on using more energy, does not invalidate well-designed programs to foster efficient energy use.

With appropriate government policies we can save energy, save money and avoid spending that money on more energy use.

- Although RE sources are diffuse, there is more than enough space on land, rooftops and coastal seas to provide for all reasonable energy demands.
- Furthermore there are enough commercially available RE technologies to convert a big enough fraction of those vast RE resources to run an industrial society.
- Most RE technologies generate the energy required to make copies of themselves within a few months to a few years. They can become breeders of RE.
- Some renewable electricity technologies are variable, while others are flexible and reliable. Combinations of variable and flexible-reliable renewable technologies can generate clean electricity with the same reliability as fossil fuels.
- There is no need to supply base-load electricity demand with base-load power stations. Combinations of flexible-reliable and variable RE technologies can easily supply base-load demand.
- Supplying the peaks in electricity demand with a 100 per cent RE system is more challenging than supplying base load. However, the challenge can be met by flexible peak-load RE power stations; commercially available storage such as hydro-electric dams, thermal storage in molten salt and stocks of renewable gases and liquids for fuelling gas turbines; geographic dispersion of wind and solar generators; and demand reduction assisted by 'smart' devices.
- It would be a waste of precious time to delay a rapid transition to 100 per cent renewable electricity in order to await cheaper batteries and new forms of energy storage.
- There is no credible scientific evidence to support claims that wind turbines can make people sick. However, there is evidence that the so-called 'wind turbine syndrome' is a nocebo effect, that is, psychologically induced by anti-wind campaigners.
- In a sustainable energy future there is a role for bioenergy produced in ecologically sustainable processes. However,

to achieve that on a large scale needs more research and development.

- Nuclear energy is too dangerous, too expensive, too slow and, in the long run, too greenhouse intensive to make a significant contribution to climate mitigation.

Fortunately the anti-sustainable-energy campaign has been less successful with the general public than the campaign of climate science denial. The rapid growth of wind and solar PV and the decline in the prices of these clean and sustainable technologies is building confidence in the community.

However, most politicians and business leaders have short time-horizons and are easily influenced by wealthy vested interests such as the fossil fuel and nuclear industries.[3] When pushed towards action by public opinion, such decision-makers tend to prefer to implement incremental changes that do not upset the status quo. Neoclassical economics, with its emphasis on small changes at the margin, reinforces this approach. The result is that, when faced with a choice between shifting from coal to cheap supplies of gas from shale and coal seams, or making the radical change to a sustainable energy system, the decision-makers are likely to choose the gas. They make this decision even though deep down they know that gas is still a greenhouse polluter and that its widespread use will still lead inevitably to climate disaster. So the principal hope for change is the community-based climate action movement.

## Proposed campaign strategy

Here are some preliminary thoughts on a campaign strategy for the climate action movement. I would welcome your comments through the book's Facebook page.[4]

As discussed in the previous chapter, the campaign must build and exert grass-roots political power to force decision-makers in government and business to implement effective policies. To build power, a military-type strategy is needed for a nonviolent transformation. Two key strategic goals are needed.

The first key goal is to identify several issues of widespread public concern that are related to climate change and/or the fossil fuel economy and devise campaign activities to engage people in addressing them. This is intended to unify the existing members of the climate action movement while, at the same time, to bring in new members and give them rewarding tasks. The issues chosen will vary from country to country and sometimes from state to state within a country. Depending upon the locality, they may include the rising price of electricity; air pollution and land degradation by mining coal, shale gas or coal seam methane; the increasing frequency of drought, wild fires, floods and intense cyclones/hurricanes/typhoons; congestion of urban roads and the lack of adequate public transport; barriers to community RE projects; and unemployment. Of course there are already campaigns on most of these issues in many countries, but they are not yet sufficiently linked together as part of national and international strategies for climate mitigation by sustainable energy.

The second key goal is to identify pressure points at which the principal barriers to change can be overcome and to apply strong pressure to these points. Some of these barriers and potential pressure points are the investments into fossil fuels by superannuation funds, banks and other financial institutions; proposed new fossil fuel developments, including coal-fired power stations, coal mines, motorways, coal seam methane, tar sands and shale oil; and press/media outlets that support climate science denial and RE denial. Another possible pressure point is sanctions against countries that persist in greatly increasing their already high GHG emissions and/or in actively undermining international processes to mitigate climate change. All the methods used against apartheid in South Africa could be contemplated, such as trade sanctions; exposure of export and import activities; diplomatic sanctions; bans and protests against sporting teams; divestment; and financial support for opposition groups that support climate action.

Different climate action groups will undertake different tactics, as discussed in Chapter 11, 'Tactics'. Some of the more experienced groups may undertake nonviolent campaigns to stop proposed fossil fuel developments. Some may campaign for divestment from fossil fuels, while others will build community RE projects. Most groups will conduct education programs for themselves and the wider community. Some of the large environmental non-government organisations will use their numbers and media connections to give weight to lobbying power-holders in government and business.

## Even bigger challenges lie ahead

While technological change can achieve rapid emission reductions, given the political will, it may not be sufficient to avert irreversible climate change and other environmental crises in the long run. We must also reform the economic system, to create a steady-state economy in which everyone has enough. And we must address the causes of population growth. Although these challenges have received little public recognition so far, there are now small but growing movements to tackle them, as discussed in Chapter 8.

Human beings can be amazingly resourceful when at long last they recognise a serious threat and an exciting opportunity. The climate crisis offers both. At heart I feel that we will, over several decades, establish a global sustainable energy system, begin the process of reforming the economic system and possibly eventually stabilise our population without coercion. Over the past 50 years, many social movements – peace, feminist, environment, social justice and consumer protection – have emerged and/or become stronger, acquiring insight and skills to challenge ruthless power structures and culturally embedded injustice, bringing about changes in attitudes and practices. The climate action movement is a growing force. I hold much hope for the future in the learning that can occur between it and other social movements.

# APPENDIX:
# WHAT YOU CAN DO

This book puts the principal emphasis on policy changes that have to be made by the three spheres of government and business to create large reductions in greenhouse gas (GHG) emissions. The most effective actions that you as an individual can take are to work through community, professional and other groups to create a mass movement to apply nonviolent pressure on governments and business to encourage them to implement the right policies and strategies. However, you shouldn't be discouraged from taking individual action; by this we can still achieve emissions reductions and accumulate valuable learning experiences for ourselves, for family, friends and neighbours.

So this appendix offers lists of actions that you as an individual or family can take to reduce your own GHG emissions. If you share your experiences you will contribute to building awareness and capacity in the wider community. Since most of the suggested actions save money as well as energy, this should make them generally socially acceptable.

The portfolio of actions varies between climate zones. This appendix is primarily for people who do not live in tropical areas.

## For tenants

- Use water-efficient showerheads in conventional high-pressure systems. (They are less effective in gravity-fed systems.) By saving hot water, you will also be saving energy.

- Put insulating blankets over indoor hot water tanks. If possible, adjust the thermostat on your hot water tank to about 67°C, which is sufficient for killing bacteria but not as wasteful of energy as higher temperatures.
- In winter, heat only the rooms that you are using. Use 'sausage'-type draught excluders under doors.
- In places where most electricity is generated from coal, plug-in gas heaters are generally responsible for much less $CO_2$ emissions than electric resistance heaters. However, they do emit air pollutants and must be used carefully. They are banned in some jurisdictions.
- In places where most electricity is generated from coal, it is preferable to limit the use of electric resistance heaters to heating a very small room for a short period of occupation. Take the usual precautions with electrical appliances in wet areas.
- An evaporative cooler is effective, except in the wet tropics, and uses much less energy than an air conditioner.
- A fan is effective everywhere for direct cooling of people.
- Compact fluorescent light bulbs can be used to replace incandescent light bulbs that are typically used for several hours per day. LED light bulbs are recommended too, but are still expensive.
- A microwave cooker is much more energy efficient than an ordinary electric cook-top and is less dangerous than forms of cooking that char the food and thus produce carcinogens.
- Purchase Green Power, if available, from an accredited energy retailer.
- Offset the GHG emissions from your trips by car or air by purchasing renewable energy (RE) offsets, if available.
- Buy a laptop computer rather than a desktop. For good posture, put the laptop on a stand and buy a separate plug-in keyboard.
- If purchasing a television, buy a liquid crystal (LCD) display rather than a plasma, which uses much more electricity.
- Use cold water, with appropriate detergent, for washing clothes.

- Use a solar clothes drier (that is, clothes line) if possible.
- Purchase electrical and gas appliances with the maximum number of stars for energy efficiency. The additional up-front costs are usually repaid within a few years.
- Clean the coils at the back of your refrigerator and ensure that they are well ventilated. Check the refrigerator's door seal and replace it if necessary.
- Switch off electrical appliances at the power point when not in use. Stand-by energy consumption, summed over several appliances and 24 hours per day, can be significant.
- Much heat is gained and lost through windows. Install blinds and/or curtains (with pelmets if the landlord permits).
- Live near public transport.
- Walk or cycle for short trips. You will become healthier as well as greenhouse friendly.
- Does your school have a 'walking bus' for escorting children to and from school? If not, consider starting one.
- Consider car-pooling or car-sharing.

## For homeowners

Homeowners can do everything that tenants can do, plus the following:
- Install insulating batts above the ceiling. If you prefer not to use fibreglass, woollen batts are available for a slightly higher price. Find out if your walls are suitable for insulating too.
- Install 'whirlybirds', wind-powered roof ventilators, to ventilate the attic space.
- For space heating, the best options in terms of $CO_2$ emissions are passive solar, active solar (not widely commercially available yet) or an *efficient* electric heat pump (reverse-cycle air conditioner), in that order of preference.
- If your roof receives sunshine for most of the year, install solar hot water. A controller that allows you to set the times when the booster switches on and off is preferable. Some local

governments require planning permission for solar hot water systems that are visible from the street.

- If your roof receives sunshine for most of the year, install a solar photovoltaic (PV) system.
- If your roof is shaded, install an electric heat pump hot water system and purchase Green Power, if available from your electricity retailer.
- Ceiling fans use much less electricity than air conditioners.
- Fluorescent lights are much more energy efficient than incandescent. For areas that require high illumination, such as workshops and kitchens, large tubular fluorescent lights fitted by an electrician are preferable to compact fluorescent bulbs.
- Cook with gas and microwave. If gas is not available, a high-efficiency electric cook-top is better than standard electric.
- If building a new house or renovating, refer to the advice in *Your Home*[1] and the book by Derek Wrigley.[2] The key goals should be to ensure that the house or extension is well-insulated, to bring as much winter sunlight as possible into the house, to exclude summer sun and to have cross-ventilation in summer.
- If you are buying a house, try to get one that is near public transport and/or your employer, shops and other services.

If some of these options are impossible for you, due to inadequate infrastructure, you can take collective action to improve matters for everyone, as discussed in Chapter 11.

When planning holidays, consider going somewhere within your own country that can be accessed by train or at least is within driving distance rather than flying distance. If you can fill the car with family and/or friends, per capita emissions will be relatively small. These are generally overlooked when holidays are planned with the help of an airline route map.

The products that we purchase all have various amounts of energy embodied in them, as the result of the raw materials they use, the manufacturing processes, their transportation and the management of associated wastes. Some guidelines[3] are to purchase, where possible, local products; reused furniture, books and toys; organic food; and ethical investment portfolios.[4] For purchases of petrol or diesel, you may wish to take into account that Exxon Mobil is funding deniers of climate science and working actively to stop government and international actions to reduce GHG emissions.

Other actions that you can take individually are:

- Write to newspapers – both major and local, magazines, newsletters, professional journals and on-line opinion – to express your support for rapid and effective climate action with sustainable energy. Put your views to talk-back radio.

- Write to your local political representatives. (This is a less effective use of an individual's time than getting a letter published.)

- Support your local environment centre, which is a valuable information resource.

- Ask your municipal, school, technical college or university library to order appropriate books and journals (see Further reading).

- Vote for a political party or candidate that has strong policies for reducing GHG emissions. If you are a member of a major federal or state political party, recommend stronger policies to support efficient energy use and RE and to stop the construction of conventional coal-fired power stations. Don't be satisfied with tokenism – the problem is too big for that.

- Join an organisation that is working to reduce emissions and thus transform your individual action into collective action, which is much more powerful.

# REFERENCES AND NOTES

All websites were accessed on 30 August 2013.

### Introduction

1 Commoner, B (1976) *The Poverty of Power: Energy and the economic crisis*, Knopf, New York, p 1.
2 Intergovernmental Panel on Climate Change (IPCC) (2013/2014 in press) *IPCC Fifth Assessment Report*, IPCC, <www.ipcc.ch>; Potsdam Institute for Climate Impact Research and Climate Analytics (2012) *Turn Down the Heat: Why a 4°C warmer world should be avoided*, World Bank, Washington DC.
3 DARA (2012) *Climate Vulnerability Monitor: A guide to the cold calculus of a hot planet*, 2nd ed, <http://daraint.org/climate-vulnerability-monitor/ climate-vulnerability-monitor-2012/report>; Gohlke, J, Thomas, R, Woodward, A et al. (2011) Estimating the global public health implications of electricity and coal consumption, *Environmental Health Perspectives*, 119: 821–26.
4 Burns, SLS (2005) *Bringing Down the Mountains: The impact of mountain top removal surface coal mining on south west Virginia communities, 1970–2004*, PhD thesis, West Virginia University.
5 Kay, D, Barbato, J, Brassington, G et al. (2006) Impacts of longwall mining to rivers and cliffs in the southern coalfield, Coal Operators Conference, <http://ro.uow.edu.au/cgi/viewcontent.cgi?article=1058&context= coal>.
6 Aleklett, K, Hook, M, Jakobsson, K et al. (2010) The peak of the oil age: Analysing the world oil production Reference Scenario in World Energy Outlook 2008, *Energy Policy*, 38: 1398–414; Association for the Study of Peak Oil and Gas <www.peakoil.net>.
7 Monbiot, G (2012) False summit, <www.monbiot.com/2012/07/02/ false-summit>. Monbiot's article is based on an uncritical acceptance of the report Maugeri, L (2012) *Oil: The next revolution,* Discussion Paper #2012/10, Harvard Kennedy School, <http://belfercenter.ksg.harvard.edu/ files/Oil-%20The%20Next%20Revolution.pdf>.
8 Severe flaws in Maugeri's argument have been identified by Sorrell, S and McGlade, C (2012) Commentary on Maugeri's decline rate assumptions, Oil Depletion Analysis Centre, <www.odac-info.org/sites/default/files/

Maugeris-decline-rate-assumptions-commentary-rev1.pdf>, based partly on Sorrell, S, Spiers, J, Bentley, R et al. (2012) Shaping the global oil peak: A review of the evidence on field sizes, *Energy*, 37: 709–24.

9  Unconventional oil and gas can be obtained from shale, tar sands, coal seams and from the conversion of coal itself into liquid fuel.

10 Aldhous, P, McKenna, P and Stier, C (2010) Gulf leak: Biggest spill may not be biggest disaster, *New Scientist*, <www.newscientist.com/special/deepwater-horizon-oil-spill>.

11 Energy Watch Group (2007) *Coal: Resources and future production*, EWG paper 1/2007, <www.energywatchgroup.org/fileadmin/global/pdf/EWG_Report_Coal_10-07-2007ms.pdf>; Mohr, SH and Evans, GM (2009) Forecasting coal production until 2100, *Fuel*, 88: 2059–67.

12 Skeptical Science <www.skepticalscience.com>.

13 Real Climate <www.realclimate.org>.

14 IPCC (2007) *Contribution of Working Groups I, II and III to the Fourth Assessment Report of the Intergovernmental Panel on Climate Change*, Synthesis Report, IPCC, Geneva.

15 Ramanathan, V and Feng, Y (2008) On avoiding dangerous anthropogenic interference with the climate system: Formidable challenges ahead, *PNAS*, 105: 14245–250.

16 Potsdam Institute (2012).

17 Meinshausen, M, Meinshausen, N, Hare, W et al. (2009) Greenhouse-gas emission targets for limiting global warming to 2°C, *Nature*, 458: 1158–62.

18 McKibben, B (2012) Global warming's terrifying new math, *Rolling Stone*, 19 July, <www.rollingstone.com/politics/news/global-warmings-terrifying-new-math-20120719>; see also Do the Math <http://math.350.org>.

19 In this book I use the term 'technology' in a broad sense to include hardware, software (how to use the hardware) and 'orgware' (organisational/institutional arrangements associated with the hardware). Land use is also included in 'technology'.

20 Pearse, G (2007) *High and Dry*, Viking, Melbourne. See also Pearse's website <www.guypearse.com>.

21 Oreskes, N and Conway, EM (2010) *Merchants of Doubt: How a handful of scientists obscured the truth on issues from tobacco smoke to global warming*, Bloomsbury Press, New York; Washington, H and Cook, J (2011) *Climate Change Denial: Heads in the sand*, Earthscan, London.

## 1. Energy and its greenhouse gas emissions

1  Watson, D (2005) *What is the Definition of Energy?*, FT Exploring Science and Technology, <www.ftexploring.com/energy/definition.html>.

2  Boyden, S and Dovers, S (1997) Humans in the biosphere. In: Diesendorf, M and Hamilton, C (eds) *Human Ecology Human Economy: Ideas for an ecologically sustainable future*, Allen & Unwin, Sydney.

3  Boyden and Dovers (1997).

4  Washington, H (2013) *Human Dependence on Nature: How to help solve the environmental crisis*, Routledge-Earthscan, London.

5   Sørensen, B (2011b) *A History of Energy: Northern Europe from Stone Age to the present day*, Earthscan, Oxford.

6   Bureau of Resources and Energy Economics (2012) *Energy in Australia 2012*, Canberra, pp 121–22, <www.bree.gov.au>.

7   Just to be confusing, they sometimes use 'power' as a generic term for electricity.

8   International Energy Agency (IEA) (2012a) *Key World Energy Statistics 2012*, Organisation for Economic Co-operation and Development (OECD)/ IEA, Paris.

9   IEA (2012a).

10  US Energy Information Administration (EIA) (2012) *Annual Energy Review 2011*, US Energy Information Administration, DOE/EIA-0384(2011), <www.eia.gov/aer>.

11  Actually oil produced from coal, tar sands and shale is more polluting in terms of GHG emissions than coal. Fortunately none of these processes is being used on a large scale and there is strong citizen resistance to such projects.

12  IEA (2012a).

13  Howarth, RW, Santoro, R and Ingraffea, A (2011) Methane and greenhouse-gas footprint of natural gas from shale formations, *Climatic Change*, 106: 679–90; Howarth, RW, Santoro, R and Ingraffea, A (2012) Venting and leaking of methane from shale gas development: Response to Cathles et al, *Climatic Change*, doi:10.1007/s10584-012-0401-0.

14  IEA (2012b) $CO_2$ *Emissions from Fuel Combustion Highlights*, OECD/IEA, Paris, Figs 5 and 6.

15  Sometimes the smaller, older black coal-fired power stations are run in this way. But brown coal power stations cannot be ramped.

16  Haslett, J and Diesendorf, M (1981) The capacity credit of wind power: A theoretical analysis, *Solar Energy*, 26: 391–401; Martin, B and Diesendorf, M (1982) Optimal mix in electricity grids containing wind power, *Electrical Power & Energy Systems*, 4(3): 155–61.

17  Boyle, G (ed) (2007) *Renewable Electricity and the Grid: The challenge of variability*, Earthscan, London.

18  World Commission on Environment and Development (1987) *Our Common Future*, Oxford University Press, New York, p 8.

19  Diesendorf, M (2000) Sustainability and sustainable development. In: Dunphy, D, Benveniste, J, Griffiths, A et al. (eds), *Sustainability: The corporate challenge of the 21st century*, Allen & Unwin, Sydney, p 23.

20  Gohlke, JM, Thomas, R, Woodward, A et al. (2011) Estimating the global public health implications of electricity and coal consumption, *Environmental Health Perspectives*, 119: 821–26; Climate and Health Alliance and Climate Institute (2012) *Our Uncashed Dividend: The health benefits of climate action*, <www.caha.org.au> and <www.climateinstitute. org.au>.

21  Byrne, J, Glover, L and Martinez, C (eds) (2002) *Environmental Justice: Discourses in international political economy*, Series on Energy and

Environment Policy, vol 8, Transaction Publishers, New Brunswick USA, Chs 1, 4, 5, 6, 7, 11.

## 2. Energy resources and technologies

1 Thomas Alva Edison in conversation with Henry Ford and Harvey Firestone (1931), quoted as a recollection of James Newton (1987) *Uncommon Friends: Life with Thomas Edison, Henry Ford, Harvey Firestone, Alexis Carrel & Charles Lindbergh*, Harcourt Inc, Orlando FL, p 31.

2 SBC Energy Institute (2013) *Carbon Capture and Storage*, Factbook, January 2013 update, Global CCS Institute, <http://cdn.globalccsinstitute.com/publications/factbook-bringing-carbon-capture-and-storage-market>.

3 Commonwealth Scientific and Industrial Research Organisation (CSIRO) (2012) *Assessing Post-Combustion Capture for Coal-fired Power stations in Asia-Pacific Partnership Countries*, Final report to Department of Resources, Energy and Tourism, EP116217.

4 International Energy Agency (IEA) (2010) *Energy Technology Perspectives 2010*, OECD/IEA, Paris, Fig ES.1.

5 Koomey, J and Hultman, NE (2007) A reactor-level analysis of busbar costs for US nuclear plants, 1970–2005, *Energy Policy*, 35: 5630–42.

6 United Nations Framework Convention on Climate Change (UNFCCC) (2008) *Identifying, Analysing and Assessing Existing and Potential New Financing Resources and Relevant Vehicles to Support the Development, Deployment, Diffusion and Transfer of Environmentally Sound Technologies*, Interim report by the chair of the Expert Group on Technology Transfer, FCCC/SB/2008/INF.7, <http://unfccc.int/resource/docs/2008/sb/eng/inf07.pdf>.

7 Foxon, TJ, Gross, R, Chase, A et al. (2005) UK innovation systems for new and renewable energy technologies: Drivers, barriers and systems failures, *Energy Policy*, 33: 2123–37.

8 Sørensen, B (2011a) *Renewable Energy*, 4th ed, Academic Press, Amsterdam.

9 ABARE and Geoscience Australia (2010) *Australian Energy Resource Assessment*, Chapter 10 <www.ga.gov.au/products/servlet/controller?event=GEOCAT_DETAILS&catno=70142>.

10 IEA-ETSAP and International Renewable Energy Agency (IRENA) (2013) *Concentrating Solar Thermal Power: Technology brief*, <www.irena.org>.

11 Solar Augmentation Project at Liddell Power Station (2013) EcoGeneration, January/February, <http://ecogeneration.com.au/news/solar_augmentation_project_at_liddell_power_station/079478>.

12 Blakers, A (2013) Explainer: What is photovoltaic solar energy?, *The Conversation*, <http://theconversation.com/explainer-what-is-photovoltaic-solar-energy-12924>.

13 European Photovoltaic Industry Association <www.epia.org/home>.

14 Actually one solar power station has been reported as generating (at a very low level) during a night of the full moon.

15 European Photovoltaic Industry Association <www.epia.org/home>.

16 Renewable Energy Policy Network for the 21st Century (REN21) (2013) *Renewables 2013 Global Status Report*, <www.ren21.net/REN21Activities/GlobalStatusReport.aspx>.

17 National Renewable Energy Laboratory, USA <www.nrel.gov/csp/solarpaces/by_country.cfm>.

18 Sørensen, B (2011b) *A History of Energy: Northern Europe from Stone Age to the present day*, Earthscan, Abingdon, Ch 10.

19 Global Wind Energy Council (GWEC) (2013) *Global Wind Statistics 2012*, GWEC, <www.gwec.net/wp-content/uploads/2013/02/GWEC-PRstats-2012_english.pdf>.

20 Greenpeace International and GWEC (2012) *Global Wind Energy Outlook 2012*, <www.gwec.net/publications/global-wind-energy-outlook/global-wind-energy-outlook-2012>.

21 GWEC (2013).

22 European Wind Energy Association (EWEA) (2011) *Wind in our Sails: The coming of Europe's off-shore wind energy industry*, EWEA, Brussels.

23 GEA (2012) *Global Energy Assessment – Toward a Sustainable Future*, Cambridge University Press, Cambridge, UK and the International Institute for Applied Systems Analysis, Laxenburg, Austria, Ch 17.

24 URS Australia (2009) *Oil Mallee Industry Development Plan for Western Australia*, URS Australia Pty Ltd for Oil Mallee Association of Western Australia Inc and the Forest Products Commission, Perth.

25 Simpson, JA, Picchi, G, Gordon, AM et al. (2009) *Short Rotation Crops for Bioenergy Systems: Environmental benefits associated with short-rotation woody crops*, IEA Bioenergy Task 30, Rotorua.

26 Stuckley, CR, Schuck, SM, Sims, REH et al. (2004) *Biomass Energy Production in Australia: Status, costs and opportunities for major technologies*, A report for the Joint Venture Agroforestry Program, RIRDC Publication no 04/031, Rural Industries Research & Development Corporation, Canberra.

27 Presnell, K (undated) The potential use of mimosa as fuel for power generation, <www.weeds.org.au/WoNS/mimosa/docs/awc15-14.pdf>.

28 IEA (2011) *Technology Roadmap: Geothermal heat and power*, OECD/IEA, Paris.

29 The 'electrical' is necessary because some geothermal plants produce only heat. In geothermal electricity the electrical energy produced is a small fraction of the thermal energy.

30 Sinden, G (2007) Characteristics of the UK wind resource: Long-term patterns and relationship to electricity demand, *Energy Policy*, 35: 112–27; see also references cited by Delucchi, MA and Jacobson, MZ (2011) Providing all global energy with wind, water, and solar power, Part II: Reliability, system and transmission costs, and policies, *Energy Policy*, 39: 1170–90.

31 Foley, AM (2012) Current methods and advances in forecasting of wind power generation, *Renewable Energy*, 37: 1–8; Mellit, A and Pavan, AM (2010) A 24-h forecast of solar irradiance using artificial neural network,

*Solar Energy*, 84: 807–21.

32  Sørensen, B (2011a), Ch 5.

33  Parkinson, G (2013) Battery storage: The numbers don't add up – yet, *RenewEconomy*, <http://reneweconomy.com.au/2013/battery-storage-the-numbers-dont-add-up-yet-97438>.

34  Parkinson, G (2013) Battery storage, take two: Why it should be a 'no brainer', <http://reneweconomy.com.au/2013/battery-storage-take-two-why-it-should-be-a-no-brainer-40883>; Muldoon, D (2013) Battery storage take 4: Skinny connections to yield fat savings, <http://reneweconomy.com.au/2013/battery-storage-take-4-skinny-connections-to-yield-fat-savings-50820>.

35  Sørensen, B and Meibom, P (2000), A global renewable energy scenario, *International Journal of Global Energy Issues*, 13: 196–276.

## 3.  Sustainable energy scenarios

1  Saint-Exupéry, A, de (1939) *Terre des Hommes (Wisdom of the Sands)*, Gallimard, Paris.

2  Stern, N (2006) *Stern Review on the Economics of Climate Change*, <www.webcitation.org/5nCeyEYJr>; Sørensen, B and Meibom, P (2000) A global renewable energy scenario, *International Journal of Global Energy Issues*, 13: 196–276; International Energy Agency (IEA) (2010) *Energy Technology Perspectives 2010: Scenarios and strategies to 2050*, OECD/IEA, Paris; Intergovernmental Panel on Climate Change (IPCC) (2011) *Special Report on Renewable Energy Sources and Climate Mitigation*, <http://srren.ipcc-wg3.de/report/IPCC_SRREN_Full_Report.pdf/view>; Jacobson, MZ and Delucchi, MA (2011) Providing all global energy with wind, water, and solar power, Part I: Technologies, energy resources, quantities and areas of infrastructure, and materials, *Energy Policy*, 39: 1154–69; Delucchi, MA, Jacobson, MZ (2011) Providing all global energy with wind, water, and solar power, Part II: Reliability, system and transmission costs, and policies, *Energy Policy*, 39: 1170–90; WBGU (German Advisory Council on Global Change) (2011) *World in Transition: A social contract for sustainability*, WBGU, Berlin; WWF (2011) *The Energy Report: 100% renewable energy by 2050*, Technical Report, World Wide Fund for Nature, <wwf.panda.org/energyreport>; GEA (2012) *Global Energy Assessment – Toward a Sustainable Future*, Cambridge University Press, Cambridge, UK and the International Institute for Applied Systems Analysis, Laxenburg, Austria, <www.globalenergyassessment.org>; Greenpeace International, EREC and GWEC (2012) *Energy [R]evolution: A sustainable world energy outlook*, 4th ed, <www.greenpeace.org/international/Global/international/publications/climate/2012/Energy%20Revolution%202012/ER2012-Briefing.pdf>; European Climate Foundation (ECF) (2010) *Roadmap 2050: A practical guide to a prosperous, low-carbon Europe*, The Hague, Netherlands, European Climate Foundation, <www.roadmap2050.eu>; Heide, D, Greiner, M, von Bremen, L et al. (2011) Reduced storage and balancing needs in a fully European power system with excess wind and solar power

generation, *Renewable Energy*, 36: 2515–23; Rasmussen, MG, Andresen, GB and Greiner, M (2012) Storage and balancing synergies in a fully or highly renewable pan-European power system, *Energy Policy*, 51: 642–51; Elliston, B, Diesendorf, M and MacGill, I (2012a) Simulations of scenarios with 100% renewable electricity in the Australian National Electricity Market, *Energy Policy*, 45: 606–13; Elliston, B, MacGill, I and Diesendorf, M (2013a) Least cost 100% renewable electricity scenarios in the Australian National Electricity Market, *Energy Policy*, 59: 270–82; AEMO (2013) *100 per cent Renewables Study: Modelling outcomes*, Australian Energy Market Operator, July; Lund, H and Mathiesen, B (2009) Energy system analysis of 100% renewable energy systems – the case of Denmark in years 2030 and 2050, *Energy*, 34: 524–31; SRU (German Advisory Council on the Environment) (2011) *Pathways Towards a 100% Renewable Electricity System*, Tech Report, Berlin, Germany, <www.umweltrat.de/EN>; Connolly, D, Lund, H, Mathiesen, BV et al. (2011) The first step towards a 100% renewable energy-system for Ireland, *Applied Energy*, 88: 502–7; Lehmann, H (2003) *Energy Rich Japan*, Technical Report, Institute for Sustainable Solutions and Innovations, <http://energyrichjapan.info>; Mason, IG, Page, SC and Williamson, AG (2010) A 100% renewable electricity generation system for New Zealand utilising hydro, wind, geothermal and biomass resources, *Energy Policy*, 38: 3973–84; Sørensen, B (2008) A renewable energy and hydrogen scenario for northern Europe, *International Journal of Energy Research*, 32: 471–500; Krajačić, G, Duić, N, Zmijarević, Z et al. (2011) How to achieve a 100% RES electricity supply for Portugal? *Applied Energy*, 88: 508–17; Kemp, M and Wexler, J (eds) (2010) *Zero Carbon Britain 2030: A new energy strategy*, Technical Report, Centre for Alternative Technology, <http://zerocarbonbritain.com>; National Renewable Energy Laboratory (NREL) (2012) *Renewable Electricity Futures*, <www.osti.gov/bridge>.

3    Kemp and Wexler (2010).

4    GEA (2012).

5    Daly, HE (1977) *Steady State Economics: The economics of biophysical equilibrium and moral growth*, WH Freeman, San Francisco; Victor, PA (2008) *Managing without Growth: Slower by design, not disaster*, Edward Elgar, Cheltenham, UK; Jackson, T (2010) *Prosperity without Growth: Economics for a finite planet*, Earthscan, London; Dietz, R and O'Neill, D (2013) *Enough is Enough: Building a sustainable economy in a world of finite resources*, BK, San Francisco.

6    IEA (2010), p 102.

7    Kemp and Wexler (2010).

8    MacKay, DJC (2009) *Sustainable Energy – without the hot air*, UIT, Cambridge, UK.

9    Desertec Foundation <www.desertec.org>.

10   NREL (2012).

11   Elliston et al. (2012a); AEMO (2013).

12   Lehmann (2003).

13   Elliston et al. (2012a).

14 Elliston, B, Diesendorf, M and MacGill, I (2012b) Reliability of 100% renewable electricity in the Australian National Electricity Market, *Proc. IRENEC 2012, 2nd International 100% Renewable Energy Conference, Istanbul, June*, pp 325–31.

15 Jacobson and Delucchi (2011).

16 Elliston et al. (2013a).

17 Riedy, C and Diesendorf, M (2003) Financial subsidies to the Australian fossil fuel industry, *Energy Policy*, 31: 125–37: Riedy, CJ (2007) *Energy and Transport Subsidies in Australia: 2007 update*, Institute for Sustainable Futures, Sydney.

18 Elliston, B, MacGill, I and Diesendorf, M (2013b) Comparing least cost scenarios for 100% renewable electricity with low emission fossil fuel scenarios in the Australian National Electricity Market, Discussion paper to be published <www.ies.unsw.edu.au/about-us/news-activities/2013/08/does-coal-have-future>.

19 NREL (2012).

20 Heide et al. (2011).

21 Rasmussen et al. (2012).

22 IEA <www.iea.org>.

23 IEA (2010), p 80.

24 Danish Government (2011) *Our Future Energy*, Copenhagen, November.

25 Scottish Government (2011) *2020 Routemap for Renewable Energy in Scotland*, Edinburgh, July; Scottish Government (2012) *2020 Renewable Routemap for Scotland – Update*, <www.scotland.gov.uk/Resource/Doc/917/0118802.pdf>.

26 NREL (2012), vol 1, p iii.

27 Danish Energy Agency (2013) Electricity consumption fell by 1.3% in 2012, 11 February, <www.ens.dk/en/info/news-danish-energy-agency/electricity-consumption-fell-13-2012>.

28 Lund, H, Hvelplund, F, Østergaard, PA et al. (2010) *Danish Wind Power Export and Cost*, CEESA Research Project, Aalborg University, <http://vbn.aau.dk/en/publications/danish-wind-power--export-and-cost(71a8b090-31b0-11df-b84e-000ea68e967b).html>.

29 Boland, J (2012) Wind power: Why is South Australia so successful?, *The Conversation*, <http://theconversation.edu.au/wind-power-why-is-south-australia-so-successful-9706>; Australian Energy Market Operator (2013) South Australian Electricity Report, 30 August, <www.aemo.com.au/Electricity/Planning/South-Australian-Advisory-Functions/South-Australian-Electricity-Report>.

30 Potsdam Institute (2012) *Turn Down the Heat: Why a 4°C warmer world should be avoided*, The World Bank, Washington DC.

31 Burger, A (2013) SW wind farms to supply clean, renewable power for less cost than natural gas, TriplePundit <www.triplepundit.com/2013/07/southwest-wind-farms-supply-clean-renewable-power-cost-natural-gas>; Paton, J (2013) Australian wind energy now cheaper than coal, gas, BNEF says, *Bloomberg New Energy Finance*, 7 February, <www.bloomberg.com/

news/2013-02-06/australia-wind-energy-cheaper-than-coal-natural-gas-bnef-says.html>.

32  ExternE studies <www.externe.info/externe_2006>.

33  Delina, L and Diesendorf, M (2013) Is wartime mobilisation a suitable policy model for rapid national climate mitigation?, *Energy Policy*, 58: 371–80.

## 4. Saving energy

1  Lovins, AB (1990) The negawatt revolution, *The Conference Board Magazine*, 27(9): 18–23, <http://thewindway.com/pdf/E90-20_NegawattRevolution.pdf>.

2  McKinsey & Company (2009) *Pathways to a Low-Carbon Economy*, version 2 of the global greenhouse gas abatement cost curve, <www.mckinsey.com/client_service/sustainability/latest_thinking/pathways_to_a_low_carbon_economy>.

3  International Energy Agency (IEA) (2012) *World Energy Outlook 2012*, OECD/IEA, Paris, p 297.

4  IEA (2012), Fig 10.14.

5  Global Energy Assessment (GEA) (2012) *Global Energy Assessment – Toward a Sustainable Future*, Cambridge University Press, Cambridge, UK and the International Institute for Applied Systems Analysis, Laxenburg, Austria, <www.globalenergyassessment.org>.

6  Strictly speaking, the term 'energy efficiency' may be applied to both efficient generation and efficient use. In this book, we only apply it to the latter.

7  See Chapter 3, 'Simulations of 100 per cent renewable electricity' and Chapter 5, 'Quantified costs and unquantified benefits' for a discussion of discount rates.

8  Rosenfeld, AH and Poskanzer, D (2009) A graph is worth a thousand gigawatt-hours: How California came to lead the United States in energy efficiency, *Innovations*, Fall issue: 57–79, Fig 7.

9  GEA (2012), Ch 10.

10  GEA (2012), Ch 10.

11  For books, you could start with Further reading. For magazines, see *Mother Earth News* <www.motherearthnews.com>;
*ReNew* <www.ata.org.au/publications/renew>;
*Home Energy* <www.homeenergy.org>.

12  GEA (2012), Ch 10.

13  GEA (2012), Ch 8.

14  Pears, A (2004) Energy efficiency – its potential: Some perspectives and experiences, Background paper for International Energy Agency Energy Efficiency Workshop, Paris, April.

15  von Weizsäcker, E, Lovins, AB and Lovins, LH (1997) *Factor 4: Doubling wealth – halving resource use*, Allen & Unwin, Sydney; Hawken, P, Lovins, A and Lovins, LH (1999) *Natural Capitalism: Creating the next industrial revolution*, Earthscan, London.

16 Hawken et al. (1999).

17 Factor 10 Institute <www.factor10-institute.org>.

18 Vine, E (2005) An international survey of the energy service company (ESCO) industry, *Energy Policy*, 33: 691–704.

19 IEA (2012).

20 European Commission, Energy efficiency, <http://ec.europa.eu/energy/efficiency/eed/eed_en.htm>.

21 Danish Energy Agency (2012) *Energy Efficiency Policies and Measures in Denmark*, October, Copenhagen.

22 US Department of Energy, Weatherization & Intergovernmental Program, <www1.eere.energy.gov/wip/wap.html>.

23 This book also has that bias, because most of my research has been in renewable energy rather than energy efficiency.

## 5. Renewable energy technology impacts

1 Chapman, S (2013) New study: Wind turbine syndrome is spread by scaremongers, *The Conversation*, 15 March, <http://theconversation.com/new-study-wind-turbine-syndrome-is-spread-by-scaremongers-12834>. Chapman is Professor of Public Health, University of Sydney.

2 Nishimura, AY, Hayashi, K, Tanaka, M et al. (2010) Life cycle assessment and evaluation of energy payback time on high-concentration photovoltaic power generation system, *Applied Energy*, 87: 2797–807.

3 Sahara Solar Breeder Foundation <www.ssb-foundation.com>.

4 Fthenakis, V (2012) How long does it take for photovoltaics to produce the energy used?, *PE magazine*, January–February: 16–17; Raugei, M, Fullana-i-Palmer, P and Fthenakis, V (2012) The energy return on investment (EROI) of photovoltaics: Methodology and comparisons with fossil fuel life cycles, *Energy Policy*, 45: 576–82; Martinez, E, Sanz, F, Pellegrini, S et al. (2009) Life cycle assessment of a multi-megawatt wind turbine, *Renewable Energy*, 34: 667–73; Desideri, DF, Zepparelli, V, Morettini, E et al. (2013) Comparative analysis of concentrating solar power and photovoltaic technologies: Technical and environmental evaluations, *Applied Energy*, 102: 765–84; Lenzen, M (2008) Life cycle energy and greenhouse gas emissions of nuclear energy: A review, *Energy Conversion and Management*, 49: 2178–99.

5 Smart, A and Aspinall, A (2009) *Water and the Electricity Generation Industry*, National Water Commission, Canberra, <www.ret.gov.au/energy/documents/sustainbility-and-climate-change/water%20and%20the%20electricity%20generation%20industry%20report.pdf>.

6 Gary, S (2010), Keeping solar panels free from dust, ABC Science, <www.abc.net.au/science/articles/2010/08/23/2988933.htm#.UcYulBa9o20>.

7 Jacobson, MZ and Delucchi, MA (2011) Providing all global energy with wind, water, and solar power, Part I: Technologies, energy resources, quantities and areas of infrastructure, and materials, *Energy Policy*, 39: 1154–69.

8 TecEco <www.tececo.com.au>.

9   Piketty, M-G, Wichert, M, Fallot, A et al. (2009) Assessing land availability to produce biomass for energy: The case of Brazilian charcoal for steel-making, *Biomass and Bioenergy*, 33: 180–90.

10  Jacobson and Delucchi (2011).

11  Hoen, B, Brown, JP, Jackson, T et al. (2013) *A Spatial Hedonic Analysis of the Effects of Wind Energy Facilities on Surrounding Property Values in the United States*, Lawrence Berkeley National Laboratory LBNL-6362E. The study covered 50 000 home sales in nine states.

12  Lago, C, Prades, A, Lechón, Y et al. (2009) *Wind Energy – the Facts: Part V Environmental issues*.

13  Intergovernmental Panel on Climate Change (IPCC), *Special Report on Renewable Energy* <http://srren.ipcc-wg3.de/report>, Ch 7.

14  Chapman, S and St George, A (2013) How the factoid of wind turbines causing 'vibroacoustic disease' came to be 'irrefutably demonstrated', *Australian & New Zealand Journal of Public Health*, 37: 244–49; and 17 reviews of the alleged 'wind turbine syndrome' cited therein.

15  Crichton, F (2013) How the power of suggestion generates wind farm symptoms, *The Conversation*, <http://theconversation.com/how-the-power-of-suggestion-generates-wind-farm-symptoms-12833>.

16  Clarke, D (2013) Who opposes wind power and why do they do it?, Friends of the Earth, Melbourne <http://yes2renewables.org/2013/07/16/who-opposes-wind-power-and-why-do-they-do-it>.

17  Keane, S (2012) The Landscape Guardians and the Waubra Foundation, *Independent Australia*, <www.independentaustralia.net/2012/environment/the-landscape-guardians-and-the-waubra-foundation>; Fowler, A (2011) Against the wind, ABC TV Four Corners, 11 July, <www.abc.net.au/4corners/content/2011/s3274758.htm>.

18  Phalan, B (2009) The social and environmental impacts of biofuels in Asia: An overview, *Applied Energy*, 86: S21–S29; Fargione, JE, Plevin, RJ and Hill, JD (2010) The ecological impact of biofuels, *Annual Review of Ecology Evolution and Systematics*, 41: 351–77.

19  Phalan (2009); Fargione et al. (2010).

20  International Renewable Energy Agency (IRENA) (2012) *Summary for Policy Makers: Renewable power generation costs*, International Renewable Energy Agency, Abu Dhabi.

21  Feldman, D, Barbose, G, Margolis, R et al. (2012) Photovoltaic pricing trends: Historical, recent and near-term projections, Technical report DOE/GO-102012-3839, November.

22  Bloomberg New Energy Finance (BNEF) (2013) Clean energy investment – Q2 2013 Fact Pack, <http://about.bnef.com/presentations/clean-energy-investment-q2-2013-fact-pack>. See also Cai, DWH, Adlakha, S, Low, SH et al. (2013) Impact of residential PV adoption on retail electricity rates, *Energy Policy*, 62: 830–43 <http://dx.doi.org/10.1016/j.enpol.2013.07.009>.

23  Lantz, E, Hand, M and Wiser R (2012) The past and future cost of wind energy, Conference paper NREL/CP-6A20-54526.

24 BNEF (2013).

25 Deutsche Welle (DW) (2012) Wind energy blowing away nuclear power, 22 April <www.dw.de/wind-energy-blowing-away-nuclear-power/a-15903703>.

26 IEA-ETSAP & IRENA (2013) *Concentrating Solar Power: Technology brief*, <www.irena.org/menu/index.aspx?mnu=Subcat&PriMenuID=36&CatID=141&SubcatID=283>.

27 Although wind farms span large areas of mostly agricultural land, they only occupy 1–3 per cent of that land. Open-cut coal mines generally occupy larger areas, for the same electricity generation. Even underground coal mines, using longwall mining, damage or put at risk quite large areas of surface land.

28 Direct employment includes those jobs created in the design, manufacturing, delivery, construction, installation, project management, and operation and maintenance of the technology concerned. Indirect employment comprises the relevant jobs of the suppliers (eg, manufacturing the glass and aluminium in a solar hot water system) and those who handle the wastes.

29 Wei, M, Patadia, S and Kammen, DM (2010) Putting renewables and energy efficiency to work: How many jobs can the clean energy industry generate in the US?, *Energy Policy*, 38: 919–31.

30 MacGill, I, Watt, M and Passey, R (2002) *The Economic Development Potential and Job Creation Potential of Renewable Energy: Australian case studies*, Commissioned by Australian Cooperative Research Centre for Renewable Energy Policy Group, Australian Ecogeneration Association and Renewable Energy Generators Association; Diesendorf, M (2004) Comparison of employment potential of the coal and wind power industries, *International Journal of Environment, Workplace and Employment*, 1: 82–90.

31 For the projected competitiveness of 100 per cent renewable electricity, with coal and gas with carbon capture and storage, see Elliston, B, MacGill, I and Diesendorf, M (2013b) Comparing least cost scenarios for 100% renewable electricity with low emission fossil fuel scenarios in the Australian National Electricity Market, Discussion paper to be published, <www.ies.unsw.edu.au/about-us/news-activities/2013/08/does-coal-have-future>.

## 6. Is nuclear energy a solution?

1 Fox, RW (presiding commissioner) (1976) *Ranger Uranium Environmental Inquiry: First Report*, AGPS, Canberra, p 185.

2 Chalmers, A (1999) *What is This Thing Called Science?*, 3rd ed, University of Queensland Press, Open University Press and Hackett, Indianapolis; Mitroff, II (1974) *The Subjective Side of Science*, Elsevier, Amsterdam.

3 My university webpage <www.ies.unsw.edu.au/our-people/associate-professor-mark-diesendorf> gives more details of research interests, publications, etc.

4  Monbiot, G (2011) Why Fukushima made me stop worrying and love nuclear power, *The Guardian*, 22 March, <www.guardian.co.uk/commentisfree/2011/mar/21/pro-nuclear-japan-fukushima>. My reply – Don't stop worrying, George – was published in *New Matilda*: <http://newmatilda.com/2011/03/23/dont-stop-worrying-george>.

5  Lovelock, J (2006) *The Revenge of Gaia: Why the Earth is fighting back – and how we can still save humanity*, Allen Lane, Santa Barbara CA.

6  Schneider, M and Froggatt, A (2013) *World Nuclear Industry Status Report 2013*, Mycle Schneider Consulting, Paris, July.

7  Schneider and Froggatt (2013).

8  A large fraction of the reactors classified as 'under construction' have been 'under construction' for over 20 years.

9  Schneider and Froggatt (2013).

10  Schneider and Froggatt (2013).

11  Mudd, GM and Diesendorf, M (2010) Uranium mining, nuclear power and sustainability – rhetoric versus reality, Sustainable Mining Conference, Kalgoorlie WA, 17–19 August. The tailing mass of 137 Mt is updated to 2012 from this paper.

12  Uranium enrichment is not required for the Canadian CANDU reactor.

13  Jowit, J and Espinoza, J (2006) Heatwave shuts down nuclear power plants, *The Guardian*, <www.guardian.co.uk/environment/2006/jul/30/energy.weather>.

14  Mortimer, N (1991) Nuclear power and global warming, *Energy Policy*, 19: 76–78.

15  Storm van Leeuwen, JW and Smith, P (2005, updated 2008) Nuclear power – the energy balance, <www.stormsmith.nl/reports.html>.

16  Lenzen, M (2008) Life-cycle energy and greenhouse gas emissions of nuclear energy: A review, *Energy Conversion & Management*, 49: 2178–99.

17  Lenzen (2008).

18  Mudd, GM (2007) Radon releases from Australian uranium mining and milling projects: Assessing the UNSCEAR approach, *Journal of Environmental Radioactivity*, 99(2): 288–315.

19  Mudd and Diesendorf (2010).

20  Lenzen, M and Munksgaard, J (2002) Energy and $CO_2$ life-cycle analyses of wind turbines – review and applications, *Renewable Energy*, 26: 339–62, Table 1.

21  ISA (Integrated Sustainability Analysis) (2006) *Life-cycle Energy Balance and Greenhouse Gas Emissions of Nuclear Energy in Australia*, Prepared for the Uranium Mining Processing and Nuclear Energy Review, Department of Prime Minister and Cabinet, Sydney, NSW, 3 November.

22  Cours des Comptes (Accounting Office of France) (2012) *The Costs of the Nuclear Power Sector*, January, Paris CEDEX, p 46.

23  The 'fast' refers to the speed of the neutrons produced. For the distinction between a fast (neutron) reactor and a fast breeder reactor, see Glossary.

24  MIT (2003) *The Future of Nuclear Power: An interdisciplinary MIT study*, <http://web.mit.edu/nuclearpower>.

25 Oak Ridge National Laboratory (2002) *Review*, vol 25, nos 3–4, Ch 4: 'By 1954, the Laboratory's chemical technologists had completed a pilot plant demonstrating the ability of the THOREX process to separate thorium, protactinium, and uranium-233 from fission products and from each other. This process could isolate uranium-233 for weapons development and also for use as fuel in the proposed thorium breeder reactors.'

26 Most of the waste would be short-lived too.

27 See ITER <www.iter.org>. The countries involved are China, European Union, India, Japan, Russia, South Korea and the USA.

28 Holdren, JP (1981) Fusion-fission hybrids: Environmental aspects and their role in hybrid rationale, *Journal of Fusion Energy*, 1(2): 197–210; Sievert, F and Johnson, D (2010) Creating Suns on Earth, *Nonproliferation Review*, 17: 323–46.

29 Sievert and Johnson (2010); Roser, D et al. (2013) Analysing the nuclear weapons proliferation risk posed by a mature fusion technology and economy, Discussion Paper, Institute of Environmental Studies, UNSW.

30 Koomey, J and Hultman, NE (2007) A reactor-level analysis of busbar costs for US nuclear plants, 1970–2005, *Energy Policy*, 35: 5630–42.

31 National Academy of Sciences (2005) *BEIR VII: Health Risks from Exposure to Low Levels of Ionising Radiation*, National Academies Press, Washington DC; UNSCEAR (2011) *Report of the United Nations Scientific Committee on the Effects of Atomic Radiation 2010*, United Nations, New York.

32 Morgan, O (2006) Nuclear costs to hit £90 bn, warns Brown, *Observer-Guardian*, 4 June, <www.guardian.co.uk/business/2006/jun/04/theobserver.observerbusiness>.

33 Nuclear Decommissioning Authority (2013) *Annual Report and Accounts 2012/2013*, p 132.

34 Sovacool, B (2011) *Contesting the Future of Nuclear Power*, World Scientific, London, pp 54–68. Sovacool is Director of the Danish Center for Energy Technology at AU Herning.

35 Chernobyl Forum (2006) *Chernobyl's Legacy: Health, environmental and socio-economic impacts*, 2nd revised ed, IAEA, Vienna.

36 Cardis, E et al. (2006) Estimates of the cancer burden in Europe from radioactive fallout from the Chernobyl accident, *International Journal of Cancer*, 119: 1224–35.

37 Yablokov, AV, Labunska, I and Bolkov, I (eds) (2006) *The Chernobyl Catastrophe: Consequences on human health*, Greenpeace, Amsterdam.

38 National Diet of Japan (2012) *Fukushima Nuclear Accident Independent Investigation Commission*, National Diet of Japan, Tokyo, p 16.

39 Sovacool, B (2013) Is Fukushima the new normal for nuclear reactors?, *The Conversation*, 27 August, <http://theconversation.com/is-fukushima-the-new-normal-for-nuclear-reactors-17391>.

40 Broinowski, R (2012) *Fallout from Fukushima*, Scribe, Melbourne.

41 Gipe, P (2012) Japan approves feed-in tariffs, *Renewable Energy World*, 22 June, <www.renewableenergyworld.com/rea/news/article/2012/06/japan-approves-feed-in-tariffs>.

42  Broinowski (2012), Ch 10.

43  Robock, A and Toon, B (2010) South Asian threat? Local nuclear war = global suffering, *Scientific American*, 302(1), January: 74–81.

44  Bryson, C (2004) *The Fluoride Deception*, Seven Stories Press, New York.

45  Silex Systems Ltd (2013) Silex completes test loop phase 1 milestone, brr media, 24 May, <www.brrmedia.com/event/112085>.

46  Von Hippel, F (2005) Presentation to UN Secretary-General's Advisory Board on Disarmament Affairs, 24 February, <www.un.org/disarmament/HomePage/AdvisoryBoard/44th_Session/PDF/NuclearFuelCycle.pdf>. The critical masses given here are based on the assumption that a neutron reflector is used in the bomb.

47  For example, 'Civil nuclear power does not produce materials usable for nuclear weapons' as claimed by then spokesman for the Australian Uranium Information Centre, Ian Hore-Lacy, in *Australasian Science* (2005), July, p 39.

48  These and many other verbatim quotations from authoritative sources are collected together with references at Canadian Coalition for Nuclear Responsibility <www.ccnr.org/Findings_plute.html>.

49  Institute for Science and International Security (ISIS) <www.isis-online.org/publications/iran/index.html>.

50  Nuclear Weapons Archive <http://nuclearweaponarchive.org/Nwfaq/Nfaq7-2.html#india>.

51  Albright, D and Hibbs, M (1992) Pakistan's bomb: Out of the closet, *Bulletin of the Atomic Scientists*, July–August: 38–43; Corera, G (2006) *Shopping for Bombs: Nuclear proliferation, global insecurity and the rise and fall of the AQ Khan Network*, Scribe, Melbourne.

52  Corera (2006).

53  Albright, D and Hinderstein, C (2004) Libya's gas centrifuge procurement: Much remains undiscovered, ISIS Online, <www.isis-online.org/publications/libya/cent_procure.html>.

54  Peter Hayes, Nautilus Institute, <http://nautilus.org/about/staff/peter-hayes/#axzz2YE94gjQ5>. Hayes is Executive Director of the Nautilus Institute for Security and Sustainability.

55  Albright, D (1994) South Africa and the affordable bomb, *Bulletin of the Atomic Scientists,* 50(4), July–August: 37–47.

56  Global Security.org <www.globalsecurity.org/wmd/world/brazil/nuke.htm>.

57  Federation of American Scientists <www.fas.org/nuke/guide/argentina/nuke/index.html>.

58  Nuclear Weapon Archive, Britain's nuclear weapons <http://nuclearweaponarchive.org/Uk/UKFacility.html>; Barnham, KWJ, Nelson, J and Stevens, RA (2000) Did civil reactors supply plutonium for weapons?, *Nature,* 407: 833–34, <www.nature.com/nature/journal/v407/n6806/full/407833c0.html>.

59  ISIS <http://isis-online.org>.

60  Kang, J, Suzuki, T and Hayes, P (2004) *South Korea's Nuclear Mis-Adventures,* Special Report: September, Nautilus Institute, <http://nautilus.org/napsnet/napsnet-special-reports/south-koreas-nuclear-mis-adventures/#axzz2drojtfGc>.

61  ISIS <http://isis-online.org>.

62  Reynolds, W (2000) *Australia's Bid for the Atomic Bomb*, Melbourne University Press, Melbourne; Broinowski, R (2003) *Fact or Fission: The truth about Australia's nuclear ambitions*, Scribe, Melbourne.

63  Albright, D and Hinderstein, C (2001) Algeria: Big deal in the desert?, *Bulletin of the Atomic Scientists*, 57(3): 45–52; Federation of American Scientists (FAS) Nuclear Information Project <www.fas.org/nuke/guide/algeria/index.html>.

64  Robock and Toon (2010).

65  Higgins, A (1994) Plutonium 'leaking' on to black market, *The Independent*, 17 August, <www.independent.co.uk/news/world/europe/plutonium-leaking-on-to-black-market-1376912.html>; Borger, J (2010) Nuclear bomb material found for sale on Georgia black market, *The Guardian*, 8 November, <www.guardian.co.uk/world/2010/nov/07/nuclear-material-black-market-georgia>.

66  MacKerron, G et al. (2006) *Economics of Nuclear Power*, A report to the Sustainable Development Commission, Paper 4 of series, 'The role of nuclear power in a low carbon economy', <www.sd-commission.org.uk/publications.php?id=339>.

67  Cooper, M (2009a) *All Risk, No Reward for Taxpayers and Ratepayers*, Institute of Energy and the Environment, Vermont Law School, November <www.vermontlaw.edu/Documents/11_03_09_Cooper%20All%20Risk%20Full%20Report.pdf>.

68  Public Citizen, *Nuclear's Fatal Flaws: Cost*, <www.citizen.org/cmep/article_redirect.cfm?ID=13449>.

69  Koplow, D (2007) *Subsidies in the US Energy Sector: Magnitude, causes and options for reform*, Earthtrack Inc, Ch 4 <www.earthtrack.net/content/subsidies-us-energy-sector-magnitude-causes-and-options-reform>.

70  Schneider, M, Thomas, S, Froggatt, A et al. (2009) *The World Nuclear Status Report 2009*, Commissioned by German Federal Ministry of Environment, Nature Conservation and Reactor Safety, p 92.

71  Meyer, B, Schmidt, S and Eidems, V (2009) *Staatliche Förderungen der Atomenergie im Zeitraum 1950–2008 (National Subsidies for Nuclear Energy 1950–2008), Study by FöS for Greenpeace*, <www.greenpeace.de/themen/atomkraft/nachrichten/artikel/atomenergie_kostet_bundesbuerger_bereits_258_milliarden_euro-1/>; reported in English by Parkin, B and van Loon, J (2009) Atomic Power's cheap-energy tag belied by aid, Greenpeace says, *Bloomberg.com*, 3 September, <www.bloomberg.com/apps/news?pid=20601100&sid=aeu3Q6i0rAlQ>.

72  WISE/NIRS (2005) Unfair Aid: The subsidies keeping nuclear energy afloat, WISE, 24 June, <www10.antenna.nl/wise/index.html?http://www10.antenna.nl/wise/630-31>.

73 Mitchell, C (2000) The England and Wales fossil fuel obligation: History and lessons, *Annual Review of Energy and the Environment*, 25: 285–311, Table 4.

74 Department of Trade and Industry (2003) *Our Energy Future: Creating a low carbon economy*, Section 4.68, <www.dti.gov.uk/files/file10719.pdf>.

75 Wintour, P (2013) Ed Davey 'will not give an inch' on nuclear power price, *The Guardian*, 6 July, <www.guardian.co.uk/politics/2013/jul/05/davey-minister-nuclear-power-hinkley-point>.

76 Cooper, M (2009b) *The Economics of Nuclear Reactors: Renaissance or relapse?*, Institute of Energy and the Environment, Vermont Law School, June <www.nonuclear.se/files/cooper200906economics_of_nuclear_reactors.pdf>.

77 Lovins, AB and Sheikh, I (2008) *The Nuclear Illusion*, Rocky Mountain Institute, Snowmass, <www.rmi.org/rmi/Library/E08-01_NuclearIllusion>.

78 MIT (2003); Keystone Center (2007) *Nuclear Power Joint Fact-Finding*, 2007, <www.keystone.org/policy-initiatives-center-for-science-a-public-policy/energy/nuclear-power-joint-fact-finding.html>; Harding, J (2007) Economics of nuclear power and proliferation risk in a carbon-constrained world, *Electricity Journal*, 20 (December): 65–76; MIT (2009) *Future of Nuclear Power: Update of MIT 2003*, <http://web.mit.edu/nuclearpower/pdf/nuclearpower-update2009.pdf>; Moody's (2008) *New Nuclear Generating Capacity: Potential credit implications for US investor owned utilities*, May; Florida Power & Light <www.fpl.com/environment/nuclear>; Severance, CA (2009) *Business Risks and Costs of New Nuclear Power*, <http://large.stanford.edu/publications/coal/references/nirs/docs/nuclearcosts2009.pdf>.

79 Sovacool (2011), p 109.

80 For US nuclear power stations, there are a wide range of construction times and hence a wide range of percentage increases in capital costs from 'overnight' to 'busbar' – see Koomey and Hultman (2007).

81 Schneider, Thomas, Froggatt et al. (2009), p 62.

82 Cooper (2009b).

83 Klein, J (2010) *Comparative Costs of California Central Station Electricity Generation*, California Energy Commission, CEC-200-2009-07SF, January.

84 Klein (2010).

85 Schneider, M and Froggatt, A (2012) *World Nuclear Industry Status Report 2012*, Mycle Schneider Consulting, Paris, p 33.

86 Anon (2011) EDF delays Flamanville 3 EPR project, *Nuclear Engineering International*, 20 July, <www.neimagazine.com/story.asp?sectioncode=132&storyCode=2060192>.

87 Kazuma Iwata, cited in NewsOnJapan.com (2001) <http://newsonjapan.com/html/newsdesk/article/89987.php>.

## 7. Transport and urban form

1 Newman, P, Beatley, T and Boyer, H (2009) *Resilient Cities: Responding to*

*peak oil and climate change*, Island Press, Washington DC.

2    Newman et al. (2009).

3    IEA (2012a) *CO₂ Emissions from Fuel Combustion: Highlights*, OECD/IEA, Paris.

4    For Australian readers, further information on air pollutants in Australia is available from the National Pollutant Inventory, <www.npi.gov.au>; State of the Environment reports, <www.deh.gov.au/soe/2001/index.html>; and the international IPCDS INCHEM website, <www.inchem.org>.

5    Bureau of Infrastructure, Transport and Regional Economics (2013) Road Deaths Australia 2012 Statistical Summary, <www.bitre.gov.au/publications/ongoing/files/RDA_Summary_2012_June.pdf>; Traffic fatalities in China dropped to 60 000 in 2012: Minister of Transport (2013) China Autoweb, <http://chinaautoweb.com/2013/06/traffic-fatalities-in-china-dropped-to-60000-in-2012-minister-of-transport>; Fewest traffic deaths seen since '51 (2013) Japan Times, <www.japantimes.co.jp/news/2013/01/05/national/fewest-traffic-deaths-seen-since-51>; US Department of Transportation (2011) State Traffic Data, <www-nrd.nhtsa.dot.gov/Pubs/811801.pdf>.

6    Newman, P and Kenworthy, J (1999) *Sustainability and Cities: Overcoming automobile dependence*, Island Press, Washington DC, p 113.

7    National Institute of Economics and Industry Research (NIEIR) (1996) *Subsidies to the Use of Natural Resources*, Section 2.6, Road Transport, A Report to Department of the Environment, Sport & Territories, Commonwealth of Australia, Canberra.

8    Data from Banfield, K, Hutabarat, R and Diesendorf, M (1999) Sydney's passenger transport: Accounting for different modes, *ATRF* (Proceedings of 23rd Australasian Transport Research Forum) 23: 269–85. The following paper corrected an error in this paper and extended the results: Diesendorf, M (2002) The effect of land costs on the economics of urban transport systems. In Wang, KCP et al. (eds) *Traffic and Transportation Studies*, Proceedings of Third International Conference on Traffic and Transportation Studies (ICTTS2002), pp 1422–29 <www.ies.unsw.edu.au/sites/all/files/TransportCosts.pdf>.

9    Centre for International Economics (2005) *Sydney's Transport Infrastructure: The real economics*, A report for the *Sydney Morning Herald*.

10   Laird, P, Newman, P and Bachels, M (2001) *Back on Track: Rethinking transport policy in Australia and New Zealand*, UNSW Press, Sydney.

11   Riedy, C and Diesendorf, M (2003) Financial subsidies to the Australian fossil fuel industry, *Energy Policy*, 31: 125–37: Riedy, CJ (2007) *Energy and Transport Subsidies in Australia: 2007 Update*, Institute for Sustainable Futures, Sydney.

12   Newman, P (1996) Urban design, transportation and greenhouse. In: Samuels, R and Prasad, D (eds) *Global Warming and the Built Environment*, E&FN Spon, London, pp 69–84.

13   Newman (1996), p 77.

14   Newman, P and Kenworthy, J (2006) Urban design to reduce automobile

dependence, *Opolis: An International Journal of Suburban and Metropolitan Studies*, 2(1), Art 3, <http://repositories.cdlib.org/cssd/opolis/vol2/iss1/art3>.

15 Newman and Kenworthy (2006).

16 Newman (1996), p 74.

17 Newman and Kenworthy (2006).

18 White, D, Sutton, P, Pears, A et al. (1978) *Seeds for Change: Creatively confronting the energy crisis*, Conservation Council of Victoria and Patchwork Press, Melbourne.

19 Victoria Transport Policy Institute, British Columbia, <www.vtpi.org>.

20 For example, Newman and Kenworthy (1999), Ch 4.

21 Newman (1996), p 81.

22 Stockholm Public Transport Plan in Brief 2020, <http://sl.se/PageFiles/838/Public_Transport_Plan_2020_in_brief.pdf>.

23 Hicks, J (2013) *The Social Context of Urban Travel Behaviour*, PhD thesis, UNSW.

24 Shahan, Z (2013) 100% electric cars outselling plug-in hybrid electric cars in the US, *RenewEconomy*, <http://reneweconomy.com.au/2013/100-electric-cars-outselling-plug-in-hybrid-electric-cars-in-us-2013>.

25 Bureau of Resources and Energy Economics (2012); Sørensen, B (2011) *Renewable Energy*, 4th ed, Academic Press, Burlington MA.

26 Sørensen (2011), p 588.

27 Foran, B and Mardon, C (1999) *Beyond 2025: Transitions to a biomass-alcohol economy using ethanol and methanol*, Working Paper Series 99/07, December, CSIRO Wildlife and Ecology, Canberra.

28 Newman, P (2012) Australia in the world scene: Some surprising trends, Presentation to Climate Action Summit, Sydney, April.

29 Puentes, R and Tomer, A (2008) *The Road … Less Traveled: An analysis of vehicle miles traveled trends in the US*, Brookings Institution, Washington DC.

30 Newman, P, Kenworthy, J and Glazebrook, G (2013) Peak car use and the rise of global rail, *Journal of Transportation Technologies*, doi:10.4236/jyys.2013.

31 Newman, Kenworthy and Glazebrook (2013).

## 8. General policies for the Great Transition

1 Senator Milne's keynote address to Solar 2013 conference, Melbourne, 24 May 2013. Milne is Leader of the Australian Greens (2013).

2 Watts, J (2009) Stern: Rich nations will have to forget about growth to stop climate change, <www.guardian.co.uk/environment/2009/sep/11/stern-economic-growth-emissions>.

3 Hamilton, C (1997) Foundations of ecological economics. In: Diesendorf, M and Hamilton, C (eds) *Human Ecology Human Economy: Ideas for an ecologically sustainable future*, Allen & Unwin, Sydney, pp 35–63, Table 2.1.

4 Mishan, EJ (1969) *The Costs of Economic Growth*, Penguin, Harmondsworth; Blatt, JM (1983) *Dynamic Economic Systems: A post-Keynesian approach*, ME Sharpe Inc, Armonk, New York, and Wheatsheaf

Books, Brighton, Sussex; Self, P (1993) *Government by the Market?: The politics of public choice*, Macmillan, London; Ormerod, P (1994) *The Death of Economics*, Faber and Faber, London; Davies, G (2004) *Economia: New economic systems to empower people and support the living world*, ABC Books, Sydney; Keen, S (2011) *Debunking Economics: The naked emperor dethroned*, 2nd ed, Zed Books, London.

5   Smith, V (2011) Power plant closures to cost US towns jobs, taxes, *The Guardian*, 20 December, <www.guardian.co.uk/world/feedarticle/10003995>.

6   Kemfert, C and Diekmann, J (2009) Emissions trading and promotion of renewable energy – we need both, *Weekly Report*, DIW Berlin (German Institute for Economic Research), 5(14), 28 May.

7   Lehmann, P and Gawel, E (2013) Why should support schemes for renewable electricity complement the EU emissions trading scheme?, *Energy Policy*, 52: 597–607.

8   Stiglitz, JE (2008) Is there a post-Washington consensus consensus? In: Serra, N and Stiglitz, JE (eds) *The Washington Consensus Reconsidered: Towards a new global governance*, Oxford University Press, Oxford, p 42.

9   Stilwell, F (2011/12) Marketising the environment, *Journal of Australian Political Economy*, no 68: 108–27.

10  Howarth, RW, Santoro, R and Ingraffea, A (2011) Methane and greenhouse-gas footprint of natural gas from shale formations, *Climatic Change*, 106: 679–90; Howarth, RW, Santoro, R and Ingraffea, A (2012) Venting and leakage of methane from shale gas development: Reply to Cathles et al, *Climatic Change*, 113: 537–49.

11  Spash, C (2010) The brave new world of carbon trading, *New Political Economy*, 15(2): 169–95; Clò, S and Vendramin, E (2012) *Is the ETS Still the Best Option?*, Istituto Bruno Leoni special report, <www.brunoleoni.it/nextpage.aspx?codice=11541>.

12  Betz, R and MacGill, I (2005) *Emissions Trading for Australia: Design, transitions and linking options*, Draft discussion paper, Centre for Energy and Environmental Markets, UNSW, <www.ceem.unsw.edu.au/publications>.

13  Regional Greenhouse Gas Initiative <www.rggi.org>.

14  Carbon pricing <www.climatechange.gov.au/international/actions/countries-acting-now/carbon-pricing>.

15  European Commission, Climate Action <http://ec.europa.eu/clima/policies/ets/index_en.htm>.

16  European Commission, Climate Action <http://ec.europa.eu/clima/policies/ets/pre2013/index_en.htm>; Clò and Vendramin (2012).

17  European Commission factsheet (undated), The EU ETS is delivering emission cuts, <http://ec.europa.eu/clima/publications/docs/factsheet_ets_emissions_en.pdf>.

18  Clò and Vendramin (2012).

19  EurActive (2012) EU reveals carbon-market reform package, <www.euractiv.com/climate-environment/eu-reveals-carbon-market-

reform-news-516023?utm_source=EurActiv%20Newsletter&utm_campaign=c6204bac22-newsletter_climate__environment&utm_medium=email>.

20  De Perthuis, C (2011) *Carbon Market Regulation: The case for a $CO_2$ central bank*, Paris Dauphin University Climate Economic Chair Information and Debates Series.

21  Muller, F (1996) Mitigating climate change: The case for energy taxes, *Environment*, 38(2): 13–20, 36–43.

22  Henry, JS (2012) *The Price of Offshore Revisited: New estimates for 'missing' global private wealth, income, inequality, and lost taxes*, Tax Justice Network, <www.taxjustice.net/cms/front_content.php?idcat=148>.

23  Stiglitz (2008). For a popular account see Stiglitz, JE (2011) Of the 1%, by the 1%, for the 1%, *Vanity Fair*, May.

24  Washington, H (2013) *Human Dependence on Nature: How to help solve the environmental crisis*, Routledge-Earthscan, London.

25  Jacobs, M (1991) *The Green Economy: Environment, sustainable development and the politics of the future*, Pluto Press, London; Victor, P (2008) *Managing without Growth: Slower by design, not disaster*, Edward Elgar, Cheltenham UK; Jackson, T (2009) *Prosperity without Growth*, Earthscan, London; Daly, HE and Farley, J (2010) *Ecological Economics: Principles and applications*, 2nd ed, Island Press, Washington DC; Dietz, R and O'Neill, D (2013) *Enough is Enough: Building a sustainable economy in a world of finite resources*, BK, San Francisco; Czech, B (2013) *Supply Shock: Economic growth at the crossroads and the steady state solution*, New Society Publishers, Gabriola Island BC.

26  Stern, N (2006) *The Stern Review on the Economics of Climate Change*, Cambridge University Press, Cambridge, p xi; Schandl, H and West, J (2012) Material flows and material productivity in China, Australia and Japan, *Journal of Industrial Ecology*, 16: 352–64.

27  Diesendorf, M (2000) Sustainability and sustainable development. In: Dunphy, D et al. (eds) *Sustainability: The Corporate Challenge of the 21st Century*, Allen & Unwin, Sydney, pp 19–37.

28  Griggs, D, Stafford-Smith, M, Gaffney, O et al. (2013) Sustainable development goals for people and planet, *Nature*, 495: 305–7.

29  Griggs et al. (2013). See UN, Millennium Development Goals and beyond, <www.un.org/millenniumgoals/>.

30  New Economics Foundation (NEF) (2009) *The Great Transition: A tale of how it turned out right*, New Economics Foundation, London, p 1.

31  Victor (2008).

32  Diesendorf, M (2010) A big Australia?, *Overland*, 201: 22–25, 29–30; O'Connor, M and Lines, WJ (2010) *Overloading Australia: How governments and media dither and deny on population*, 4th ed, Envirobook, Sydney.

33  Chambers, R (1983) *Rural Development: Putting the last first*, Wiley, New York.

34  UNFCCC <http://unfccc.int/essential_background/items/6031.php>.

35  UNFCCC website.

36  Schiermeier, Q (2012) The Kyoto Protocol: Hot air, *Nature*, 491: 656–58.

37  Clemons, EK and Schimmelbusch, H (2007) The environmental prisoners' dilemma, the green dilemma, <http://opim.wharton.upenn.edu/~clemons/blogs/prisonersblog.pdf>.

38  Global Commons Institute <www.gci.org.uk/contconv/cc.html>.

39  Hamilton, C (2013) *Earthmasters: The dawn of the age of climate engineering*, Allen & Unwin, Sydney.

## 9. Targeted policies for renewable energy

1   Green, MA (2002) Acceptance speech, Right Livelihood Award, <www.rightlivelihood.org/speech_green.html>. Green is Scientia Professor and Director of the Centre for Advanced Photovoltaics at UNSW.

2   However, this situation offers an opportunity to a good screenwriter.

3   In this chapter 'technologies' is used in a broad sense to include measures such as behavioural change that do not need hardware.

4   Foxon, TJ, Gross, R, Chase, A et al. (2005) UK innovation systems for new and renewable energy technologies: Drivers, barriers and systems failures, *Energy Policy*, 33: 2123–37.

5   See EU Climate Action, What is the EU doing about climate change?, <http://ec.europa.eu/clima/policies/brief/eu/index_en.htm>.

6   European Commission (2013) *Renewable Energy Progress Report*, Report from the Commission to the European Parliament, the Council, the European Economic and Social Committee and the Committee of the Regions, <http://eur-lex.europa.eu/LexUriServ/LexUriServ.do?uri=COM:2013:0175:FIN:EN:PDF>.

7   Keppley, JM (2012) A comparative analysis of California and German renewable energy policy, *Josef Korbel Journal of Advanced International Studies*, 4 (summer); Grueneich, DN (2011) California State Policy on Sustainable Energy, *AIP Conf. Proc*, 1401:153–61, doi:10.1063/1.3653849.

8   See Paul Gipe's website <www.wind-works.org/cms>.

9   Couture, TD and Bechberger, M (2013) Pain in Spain: New retroactive changes hinder renewable energy, *Renewable Energy News*, 19 April, <www.renewableenergyworld.com/rea/news/article/2013/04/pain-in-spain-new-retroactive-changes-hinders-renewable-energy>.

10  Fripp, M (2008) European experience with tradable green certificates and feed-in tariffs for renewable electricity support, Report to FPL Energy, <www.academia.edu/1017216/European_Experience_with_Tradable_Green_Certificates_and_Feed-In_Tariffs_for_Renewable_Electricity_Support>.

11  Buckman, G and Diesendorf, M (2010) Design limitations in Australian renewable energy policies, *Energy Policy*, 38: 3365–76; addendum 38: 7539–40.

12  Buckman and Diesendorf (2010).

13  Haas, R, Resch, G, Panzer, C et al. (2011) Efficiency and effectiveness of promotion systems for electricity generation from renewable energy sources – Lesson from EU countries, *Energy*, 36: 2186–93.

14 Pfund, N and Healey, B (2011) *What Would Jefferson Do? The historical role of federal subsidies in shaping America's energy future*, <www.dblinvestors. com/documents/What-Would-Jefferson-Do-Final-Version.pdf>.

15 Riedy, C and Diesendorf, M (2003) Financial subsidies to the Australian fossil fuel industry, *Energy Policy*, 31: 125–37; Riedy, C (2007) *Energy and Transport Subsidies in Australia: 2007 Update*, Institute for Sustainable Futures, UTS, Sydney.

16 ExternE studies <www.externe.info/externe_d7>.

17 International Energy Agency (IEA) (2012) *World Energy Outlook 2012*, OECD/IEA, Paris, pp 1, 69–71. For more detail on the method, see <www.worldenergyoutlook.org/resources/energysubsidies>.

18 Riedy and Diesendorf (2003); Riedy (2007).

19 Biswas, W, Diesendorf, M and Bryce, P (2004) Can photovoltaic technologies help attain sustainable rural development in Bangladesh?, *Energy Policy*, 32: 1199–207.

20 The IEA's New Policies Scenario is business-as-usual plus policies that have been announced but not yet implemented. The 450 Scenario would limit $CO_2$ concentration in the atmosphere to 450 ppm with a probability of 45 per cent.

21 IEA (2012), pp 252–53.

22 Clean Energy Finance Corporation <www.cleanenergyfinancecorp.com. au>.

23 Climate Bonds Initiative <http://climatebonds.net/about>.

24 Climate Bonds Initiative (2013) *Bonds and Climate Change: The state of the market in 2013*, <http://climatebonds.net>.

25 Solar Mosaic <https://joinmosaic.com/how-it-works>.

26 Kennedy, D (2012) *Rooftop Revolution*, Berrett-Koehler, San Francisco; see also Sungevity <www.sungevity.com>.

27 However, China has initiated a big construction program of ultra-high voltage transmission lines to bring wind power from the west to demand centres in the east.

28 European Wind Energy Association (2011) *Offshore Electricity Grid Infrastructure in Europe*, <www.ewea.org/fileadmin/ewea_documents/ documents/publications/reports/OffshoreGrid_report.pdf>.

29 Atlantic Wind Connection <http://atlanticwindconnection.com/news>.

30 Desertec Foundation <www.desertec.org>.

31 Department of Energy and Climate Change (2013) The heat is on for householders, <www.gov.uk/government/news/the-heat-is-on-for-householders>.

32 Munksgaard, J and Morthorst, PE (2008) Wind power in the Danish liberalised power market – policy measures, price impact and investor incentives, *Energy Policy*, 36: 3940–47.

33 Sensfuss, F, Ragwitz, M and Genoese, M (2008) The merit-order effect: A detailed analysis of the price effect of renewable electricity generation on spot market prices in Germany, *Energy Policy*, 36: 3086–94; Weigt, H (2009) Germany's wind energy: The potential for fossil capacity replacement and

cost saving, *Applied Energy*, 86: 1857–63.

34 Sáenz de Miera, G, Gonzalez, PdR and Vizcaino, I (2008) Analysing the impact of renewable energy support schemes on power prices: The case of wind energy in Spain, *Energy Policy*, 36: 3345–59.

35 Ray, S, Munksgaard, J, Morthorst, PE et al. (2010) Wind energy and electricity prices, exploring the merit order effect, a literature review by Póyry for the European Wind Energy Association, <www.ewea.org/fileadmin/ewea_documents/documents/publications/reports/MeritOrder.pdf>.

36 Woo, C, Horowitz, J, Moore, J et al. (2011) The impact of wind generation on the electricity spot-market price level and variance: The Texas experience, *Energy Policy*, 39: 3939–44.

37 Cutler, NJ, Boerema, ND, MacGill, IF et al. (2011) High penetration wind generation impacts on spot prices in the Australian national electricity market, *Energy Policy*, 39: 5939–49; McConnell, D, Hearps, P, Eales, D et al. (2013) Retrospective modelling of the merit order effect on wholesale electricity prices from distributed photovoltaic generation in the Australian electricity market, *Energy Policy*, 58: 17–27.

38 Elliston, B, MacGill, I and Diesendorf, M (2013b) Comparing least cost scenarios for 100% renewable electricity with low emission fossil fuel scenarios in the Australian National Electricity Market, Discussion paper to be published <www.ies.unsw.edu.au/about-us/news-activities/2013/08/does-coal-have-future>.

39 The 'capacity credit' of wind power can be defined to be the extent to which wind power in an electricity grid can substitute for the generating capacity of conventional power stations. For small penetrations of wind power into the grid, the capacity credit of wind power is approximately equal to the average wind power. However, for large penetrations capacity credit tends to a constant, which is proportional to the probability of zero wind power at the site. See Haslett, J and Diesendorf, M (1981) The capacity credit of wind power: A theoretical analysis, *Solar Energy*, 26: 391–401.

40 IEA (2012) *World Energy Outlook 2012*, OCED/IEA, Paris, p 1.

41 Smith, R (2013) US electricity use on wane, *Wall Street Journal*, 2 January.

42 Brazzale, R (2013) Electricity demand's speedy descent, *Climate Spectator*, 8 January, <www.businessspectator.com.au/article/2013/1/8/smart-energy/electricity-demands-speedy-descent>.

43 Independent Competition and Regulatory Commission (2011) Final decision, retail prices for non-contestable electricity customers 2011–2012, p 11, <www.icrc.act.gov.au/energy/electricity>.

44 Parkinson, G (2013) Energy death spiral – consume more, or prices will rise, *RenewEconomy*, 25 March <http://reneweconomy.com.au/2013/energy-death-spiral-consume-more-or-prices-will-rise-90603>.

45 Couture and Bechberger (2013); Parkinson, G (2012) Germany slashes solar tariffs, *RenewEconomy*, 24 February, <http://reneweconomy.com.au/2012/mixed-greens-germany-slashes-solar-tariffs-88470>; Harvey, F (2013) Solar

companies to sue UK government for £140m over feed-in tariff cuts, *The Guardian*, 23 January, <www.guardian.co.uk/environment/2013/jan/23/solar-companies-feed-in-tariff-cuts>; Hurst, D (2012) Solar shake-up slashes feed-in tariff, *Brisbane Times*, 25 June, <www.brisbanetimes.com.au/queensland/solar-shakeup-slashes-feedin-tariff-20120625-20y6c.html>.

46  e-Lab (2013) *A Review of Solar PV Benefit and Cost Studies*, Rocky Mountain Institute, Boulder CO, <www.rmi.org/elab_emPower>.

47  Australian Government (2012) *Energy White Paper 2012*, Department of Resources, Energy & Tourism, Canberra, p 184. (The calculation is from Queensland, Australia, in 2011.)

48  Pears, A (2013a) Energy revolution or bloody war – it's our choice, *The Conversation*, 20 April, <http://theconversation.com/energy-revolution-or-bloody-war-its-our-choice-13294>. Such transfers would be subject to reasonable and standardised safety and power system stability requirements.

49  Pears, A (2013b) If I ran an electricity network, *Climate Spectator*, 27 March, <www.businessspectator.com.au/article/2013/3/27/energy-markets/if-i-ran-electricity-network…>.

50  Pears (2013a).

## 10. Who will drive the transition?

1  Herrmann, D (1999) *Helen Keller: A life*, University of Chicago Press, Chicago, p 222.

2  Beder, S (2006) *Suiting Themselves: How corporations drive the global agenda*, Earthscan, London.

3  Beder, S (2002) *Global Spin: The corporate assault on environmentalism*, Green Books, Totnes.

4  Weinstein, D (1979) *Bureaucratic Opposition: Challenging abuses at the workplace*, Pergamon Press, New York.

5  Anderson, RC (1998) *Mid-Course Correction: Toward a sustainable enterprise: the Interface model*, Peregrinzilla Press, Atlanta, GA.

6  In one version of this story, it is claimed that the knowledge is transferred to monkeys on other islands by a kind of telepathic means. Our use of the 100 monkeys example ignores this incredible version.

7  International Renewable Energy Agency (IRENA) (2013) *30 Years of Policies for Wind Energy: Lessons from 12 wind energy markets*, Ch 5 'Denmark', <www.irena.org/menu/index.aspx?mnu=Subcat&PriMenuID=36&CatID=141&SubcatID=281>.

8  Energiegenossenschaften: Ergebnisse der Umfrage des DGRV und seiner Mitgliedsverbände im Frühsommer (2012), <www.genossenschaften.de/sites/default/files/Auswertung%20Umfrage%20Energiegenossenschaften.pdf>.

9  Potsdamer Solarverein <www.potsdamer-solarverein.de>.

10  Güssing as a Model for Regional Economic Improvement <www.youtube.com/watch?v=H1WsbQQNsV0>.

11  Danish Island 'CO$_2$ Negative' <www.youtube.com/watch?v=1_ZY0ilFdYw>.

12  BürgerEnergie Berlin eG i.G. <www.buerger-energie-berlin.de>.

13  *The Local*, Germany edition, <www.thelocal.de/money/20121111-46101. html#.UXnw4b9y7aq>.

14  National Rural Electric Cooperative Association <www.nreca.coop>.

15  National Renewables Cooperative Organization <www.nrco.coop>.

16  Oreskes, N and Conway, EM (2010) *Merchants of Doubt: How a handful of scientists obscured the truth on issues from tobacco smoke to global warming*, Bloomsbury Press, New York; Washington, H and Cook, J (2011) *Climate Change Denial: Heads in the sand*, Earthscan, London.

17  Center for Media and Democracy, Global Climate Coalition, Sourcewatch <www.sourcewatch.org/index.php/Global_Climate_Coalition>.

18  Goldenberg, S (2012) Conservative thinktanks step up attacks against Obama's clean energy strategy, *The Guardian*, 8 May, <www.guardian. co.uk/environment/2012/may/08/conservative-thinktanks-obama-energy-plans?newsfeed=true>.

19  Walmart, Greenhouse Gas Emissions, <http://corporate.walmart. com/global-responsibility/environment-sustainability/greenhouse-gas-emissions>.

20  Google, Our footprint: Beyond zero, <www.google.com.au/green/ bigpicture/#beyondzero-footprint>.

21  Even in Germany large corporations, generally those with the highest electricity demands, are exempt from paying the EEG (Renewable Energy Act) charge for feed-in tariffs.

## 11. Citizen action

1  Alinsky, SD (1971) *Rules for Radicals*, Random House, New York, p 113.

2  Alinsky (1971), p 113.

3  Sharp, G (1973) *The Politics of Nonviolent Action*, Porter Sargent, Boston, p 3. See also his website with video at <http://howtostartarevolutionfilm. com/index.php/about/gene-sharp>.

4  The Right Livelihood Award <www.rightlivelihood.org>.

5  Alinsky (1971).

6  Obama, B (1995) *Dreams from my Father*, Crown, New York, Chs 7–14.

7  Sharp, G (2003) *There Are Realistic Alternatives*. Albert Einstein Institute, Boston, Ch 2: The Importance of Strategic Planning in Nonviolent Struggle, <www.aeinstein.org/wp-content/uploads/2013/09/TARA.pdf>.

8  Bobo, K, Kendall, J and Max, S (2001) *Organising for Social Change: Midwest Academy manual for activists*, 3rd ed, Seven Locks Press, Washington.

9  This list is modified slightly from that of Bobo et al (2001).

10  Robinson, L (2012) *Changeology: How to enable groups, communities, and societies to do things they've never done before*, Scribe, Melbourne.

11  Bellamy, D (2005) Glaciers are cool, *New Scientist*, 2495, 16 April; refuted by Monbiot, G (2005) Junk science, *The Guardian*, 10 May, <www.monbiot. com/archives/2005/05/10/junk-science>.

12  Lakoff, G (2004) *Don't Think Like an Elephant!: Know your values and frame the debate*, Chelsea Green, White River Junction VT.

13  Beder, S (2000).

14  King, ML Jr (1964) *Why Can't We Wait*, Signet, New York, p 12.

15  Gandhi, MK (1927, 1929) *An Autobiography or The Story of my Experiments with Truth*, Navajivan Publishing House, Ahmedabad; Merton, T (1964) *Gandhi on Non-Violence*, New Directions, New York.

16  There are many documented cases, eg, Churchill, W and Vander Wall, J (1988, 2002) *Agents of Repression: The FBI's secret wars against the black panther party and the American Indian movement*, South End Press, Boston; Ganser, D (2004) *NATO's Secret Armies: Operation Gladio and terrorism in Western Europe*, Routledge, New York.

17  Coover, V, Deacon, E, Esser, C et al. (1978) *Resource Manual for a Living Revolution*, 2nd ed, Movement for a New Society, Philadelphia.

18  These actions could be unlawful if unauthorised. Check the law in your jurisdiction before proceeding.

19  Adam, D (2006) Royal Society tells Exxon: Stop funding climate change denial, *The Guardian*, 20 September, <http://environment.guardian.co.uk/print/0,,329580967-121568,00.html>.

20  Legal advice should be sought before undertaking this tactic.

21  Fossil Free <http://gofossilfree.org/about>.

22  McKibben, B (2013) To my foreigner's eye, Address to National Press Club, Canberra, 6 June, <http://350.org/en/about/blogs/my-foreigners-eye>.

23  Alinsky (1971), pp 170–83.

24  A more extensive list is provided at the end of this book.

25  For more details see the Appendix.

## 12. Conclusion

1  Bakker, P (2013) University of Cambridge Annual Distinguished Lecture Series in Sustainable Development. Bakker is President, World Business Council for Sustainable Development.

2  Frankfurt School-UNEP Centre/BNEF (2013) *Global Trends in Renewable Energy Investment 2013*, Frankfurt School of Finance and Management, Frankfurt am Main <www.fs-unep-centre.org>.

3  Actually, they are almost the same industry, because some of the biggest multinational corporations in fossil fuels are also among the biggest uranium miners (eg, BHP Billiton and Rio Tinto).

4  This book's Facebook page is http://www.facebook.com/SustainableEnergy SolutionsForClimateChange.

## Appendix: What you can do

1  *Your Home*: a suite of consumer and technical materials and tools developed to encourage the design, construction or renovation of homes to be more environmentally sustainable, <www.yourhome.gov.au>.

2  Wrigley, D (2005) *Making Your Home Sustainable*, Scribe, Melbourne.

3  See Green Home <www.acfonline.org.au/greenhome>.

4  Note that some 'ethical' investment products permit uranium mining and nuclear power, some even permit fossil fuels, so choose carefully.

# GLOSSARY

The following definitions are given within the contexts of climate mitigation or
    energy.

**abatement** See **mitigation**.

**adaptation** Technical term for reducing the impacts of climate change rather
    than the causes. Eg, building sea-walls; shifting farms from drought areas to
    those with higher rainfall; spraying artificial snow onto ski slopes; improving
    medical and public health facilities for treating the spread of malaria and
    dengue.

**affluence** Consumption (household expenditure) per person.

**anthropogenic** Caused by humans.

**astroturfing** Formation of front organisations by vested interests.

**automobile city** City in which most passenger transport is by automobile.

**baseline scenario** Scenario created for the purpose of comparing other scenarios.
    It may or may not be a business-as-usual scenario.

**base-load demand** Minimum daily level of demand for electric power on the
    grid.

**base-load power station** Power station that operates at rated power 24 hours per
    day, 7 days a week, wherever possible.

**biochar** Organic charcoal formed from burning biomass in reduced oxygen.

**biodiesel** Substitute for diesel fuel produced from vegetable oils or tallow.

**bioelectricity** Electricity derived from biomass.

**bioenergy** Energy derived from biomass.

**biofuel** Fuel derived from biomass. Some usages limit it to liquid fuel from
    biomass.

**biomass** Recent organic material, either plant or animal. Its stored solar energy
    may be converted into useful energy either by direct combustion or by
    first converting it into more useful forms by such processes as gasification,
    fermentation/distillation, anaerobic digestion and pyrolysis.

**biosequestration** Capture of $CO_2$ by plants by means of photosynthesis and
    the resulting storage of $CO_2$, for different periods of time, in the plants
    themselves, leaf litter and soil.

**business-as-usual scenario** Energy use scenario that starts from the present
    pattern of energy use and places no environmental constraints on future
    economic activity or technology choice.

**cap-and-trade** Emissions trading scheme that places firm limits or 'caps' on total
    emissions in future years.

**capacity credit** (eg, of wind power) Extent to which wind power in an electricity grid can substitute for the generating capacity of conventional power stations.

**capacity factor** (of a power station) Average power output divided by rated power, usually expressed as a percentage.

**carbon capture and storage/sequestration (CCS)** Capture of $CO_2$ from large point sources of emission and its compression and injection into storages, such as underground geological formations, or oceans, or vegetation, or industrial processes. (Compare **geosequestration**.)

**carbon emissions** Shorthand for $CO_2$-equivalent greenhouse gas emissions. Includes emissions from all human-induced greenhouse gases, including those that do not contain carbon.

**carbon price** Generic term for an economic value placed on the emission of greenhouse gases into the atmosphere due to human activity; most, but not all, of the gases contain carbon atoms.

**carbon tax** Tax on a fossil fuel that is proportional to the fuel's carbon content.

**champion** High-profile person who is identified with a cause, eg, Al Gore, James Hansen. The champion may or may not also be an organiser.

**climate action group** Non-government organisation devoted wholly to reducing greenhouse gas emissions.

**climate action movement** Collection of non-government organisations and individuals working to reduce greenhouse gas emissions.

**$CO_2$-equivalent emissions** A means of comparing the global warming potentials of all greenhouse gases with $CO_2$.

**coal seam gas** See **coal seam methane**.

**coal seam methane** Methane gas found in coal mines; chemically the same as natural gas.

**co-firing** Burning fossil fuels and biomass together.

**cogeneration** See **combined heat and power**.

**combined cycle** Type of power station with two stages. Waste heat from the first stage, a gas turbine, is used to produce steam for the second stage, a boiler driving a steam turbine. Fuel can be gas, coal or biomass; however, in current practice is usually gas.

**combined heat and power** Power station in which the waste heat is utilised either for heating water, space heating or industrial process heat.

**competitive market** Textbook definition is a market in which none of the buyers or sellers can influence prices. But, in terms of comparing with the real world, it might be better to define competitive market as an idealised market, which does not suffer market failure.

**complementary measures** Policy instruments other than a carbon price for reducing emissions. These include tradable renewable energy certificates, feed-in tariffs and tax concessions.

**concentrated solar thermal power** See **solar thermal electricity**.

**Contraction and Convergence** Process in which an international greenhouse target, below the existing level of emissions, is achieved by developed countries reducing their per capita emissions and developing countries increasing their per capita emissions, until every country converges to the same average per capita emissions.

**cost curve** Graph that ranks abatement options according to cost, while showing the quantity of abatement that could be achieved by each option at its own cost level.

**curtailed energy** See **spilled energy**.

**degression** Planned reduction of a feed-in tariff as the costs of the technology covered by the tariff decline.

**demand management** Deliberately changing the demand for energy by, eg, reducing peaks, or increasing troughs, or decreasing daily energy demand. May or may not involve increasing the efficiency of energy use.

**denier** Someone who denies a large body of scientific evidence from climate science.

**Direct Normal Irradiance** or **Incidence** Solar radiation that comes in a straight line from the direction of the Sun at its current position in the sky.

**discount rate** Interest rate used to discount income or expenditure in the future (see **net present value**) due to preference for consumption now rather than later. It is often expressed in 'real' terms, that is, with the rate of inflation subtracted.

**dispatch** To connect an operating power station to the network.

**distribution line** Power line for local distribution of electricity (eg, in suburbs) at low voltages.

**economically efficient** Less expensive for the same outcome.

**efficiency of energy conversion** Useful energy output divided by energy input. Usually expressed as a percentage by multiplying by 100.

**efficient energy use** (often shortened to **energy efficiency**) Using less energy to provide the same energy services.

**embodied energy** Energy inputs to a technology over its whole life-cycle.

**emissions-intensive trade-exposed industries** Industries with high greenhouse gas emissions per unit of production that must compete in international markets.

**emissions trading scheme** Scheme in which tradable permits to emit pollutants are allocated or auctioned to emitters.

**end-use energy** Energy at the point of use by 'consumers'.

**energy** Capacity to do mechanical work, measured in joules.

**energy conservation** Reducing the number or quantity of energy services.

**energy efficiency** See **efficient energy use**.

**energy payback period** Time taken for an energy supply system to generate the full life-cycle energy requirement for constructing, fuelling and decommissioning itself.

**energy performance contractor** See **energy service company**.

**energy productivity** See **efficient energy use**.

**energy return on investment** Lifetime energy output of a technology divided by its life-cycle energy inputs.

**energy service** Task or service that involves energy as an input: eg, home heating, office lighting, transportation. The focus is on the service rather than the quantity and type of energy supplied. To provide the energy service with low $CO_2$ emissions usually involves energy efficiency as well as energy supply.

**energy service company** Business selling energy services rather than just particular forms of energy. Sometimes called 'energy performance contractor'.

**entropy** Measure of disorder in a physical system.

**Erneuerbare-Energien-Gesetz** Germany's Renewable Energy Act.

**externality** or **external cost** Something which affects a buyer's or seller's utility or profit which is not included in the prices of goods and services exchanged in the market of interest: eg, the environmental and health costs of burning coal.

**envelope** (of a building) Elements of a building forming the boundary between the indoor and outdoor environments, comprising roof, external walls, windows and floor.

**fast breeder reactor** Type of fast neutron reactor that produces more fissile material than it consumes. The plutonium breeder has, around the core, non-fissile U-238, which is partially converted into Pu-239 by exposure to fast neutrons.

**fast neutron reactor** Type of nuclear reactor in which the fission chain reaction is sustained by fast neutrons. Such a reactor needs no neutron moderator, but must use fuel that is relatively rich in fissile material compared with an ordinary 'burner' reactor, that is, plutonium or relatively highly enriched uranium. Some, but not all, fast neutron reactors are fast breeder reactors.

**feedback** or **amplification process** When the output of a process increases (positive feedback) or decreases (negative feedback) the input: eg, if global warming melts glaciers to the extent that the Earth reflects less sunlight back into space, then global warming will be increased due to positive feedback.

**feed-in tariff** Premium long-term electricity tariff paid by a utility for electricity that it must purchase, which is fed back into the grid from a renewable energy source. The price is guaranteed by the government and usually paid for by electricity consumers, while the quantity of electricity sold is determined by the market.

**fissile** (adjective) Element whose atomic nucleus is capable of undergoing nuclear fission (splitting), as the result of being struck by a neutron.

**fission** (nuclear) Splitting of a heavy atomic nucleus into two smaller nuclei with the accompanying release of energy.

**flexible** Power station that can be turned on and off rapidly, and its output can be varied rapidly, when required.

**forcing** Imposed perturbation of Earth's energy balance. It can be natural, such as a large volcanic eruption that increases the fine particles in the lower stratosphere reflecting sunlight to space and thus reducing the solar energy delivered to Earth's surface. Or it can be human-made, such as the emission of greenhouse gases by fossil-fuelled power stations and motor vehicles.

**fossil fuel** Coal, oil or gas.

**frame** (verb) To present an issue to the public in the context and language that you choose.

**fuel cell** Electrochemical cell in which the energy of a reaction between a fuel, such as hydrogen, and an oxidant, such as oxygen, is converted directly and continuously into electricity.

**fugitive emissions** Emissions from fossil fuels other than those resulting directly from combustion. They include methane gas emitted from coal mines and leaky gas pipelines, and $CO_2$ and methane vented from gas fields.

**fusion (nuclear)** Fusion of the nuclei of light elements to form heavier elements with the release of energy.

**gas** In this book, 'gas' denotes either natural gas or coal seam methane or shale gas or renewable gases such as biogas and syngas. It does *not* denote gasoline (petrol) or any other liquid.

**geosequestration** Particular case of carbon capture and storage/sequestration (CCS) involving storage of $CO_2$ in underground geological formations, such as depleted oil and gas wells, salty aquifers, and deep unminable coal seams, at depths of at least 800 metres.

**Global Horizontal Irradiance** or **Incidence** Total amount of sunshine received from above by a horizontal surface.

**global warming potential** Ratio of global warming from 1 kg of a greenhouse gas to 1 kg of carbon dioxide over 100 years.

**grandfathering** Allocating emission permits free of charge to emitting industries in proportion to their current or recent emissions.

**greenhouse-intensive** Adjective describing a task or service that is high in greenhouse gas emissions per unit of output.

**Greenhouse Mafia** Leading greenhouse-gas-emitting industries in Australia applied this term to themselves, in the context of these industries' influence upon government policy.

**Green Power** Scheme in which electricity retailers provide customers with electricity from certified renewable sources for an additional charge.

**grid** Network of transmission lines joining a number of power stations to the main sites of electricity use.

**heat pump** A device that transfers heat from a colder area to a warmer area.

**high-grade uranium ore** Ore containing at least 0.1 per cent uranium oxide, $U_3O_8$.

**human ecology** Study of the interactions of humans with other species and their non-living environment.

**identity** (mathematical) Equation that is true by definition.

**insolation** Solar radiation; or more specifically, the rate of delivery of solar radiation per unit area of horizontal surface.

**integral fast reactor** Fast neutron reactor that has a particular type of on-site reprocessing plant.

**integrated gasification combined cycle** Combined cycle power station that uses gasified coal (or gasified biomass) as a fuel.

**integrated least-cost energy planning** Planning and delivery of energy services at least cost to society. Sometimes called 'integrated resource planning', or just 'least-cost planning'.

**intermediate load** Section of power demand between base load and peak load.

Also refers to power stations that supply intermediate-load demand.

**intermittent** Random variation (of a power source).

**isotopes** Chemically identical elements that differ slightly in atomic weight and other physical properties (eg, whether they are fissile).

**job-year** The equivalent of one full-time job for one year.

**kinetic energy** Energy of motion.

**ligno-cellulose** Essential part of woody cell walls of plants, consisting of cellulose associated with lignin.

**load** Demand for electricity.

**local centre** Part of a city with higher than average residential and job density, having a radius of about 1 kilometre and containing up to about 10 000 people and jobs.

**loss-of-load probability** Average number of hours per year that electricity supply fails to reach demand.

**low-grade uranium ore** Ore containing 0.01 per cent or less uranium oxide, $U_3O_8$.

**macroeconomic** Referring to an economy as a whole, or its major components, as opposed to individual industries, firms, or households.

**market failure** Conditions of perfect competition do not apply in a market. Can arise because some buyers or sellers can influence prices, or **externalities**, or 'public goods' play a significant role, or there is insufficient information, or institutional barriers to market operation.

**merit order** Priority order for dispatching power stations into the grid. A power station that is 'top of the merit order' is dispatched first.

**mitigation** Abatement. Both are technical terms for reducing greenhouse gas emissions.

**model** Simplified representation of a real system.

**negawatt** Power saved by demand reduction measures.

**neoclassical economics** Focuses on how idealised markets determine prices, outputs, and incomes. (For more information see Box 8.1.)

**net present value** The sum of all incoming cash flows minus the sum of all outgoing cash flows, where each cash flow in the future is discounted.

**nocebo effect** Medical symptoms that occur after receiving treatment when the treatment is inert or a sham.

**non-government organisation (NGO)** Community-based organisation. The definition varies, however as used in this book it excludes big businesses as well as government agencies and includes small businesses and trade unions.

**nuclear winter** One of the impacts of nuclear war, in which vast quantities of soot are lofted into the stratosphere and screen the Sun for several years, decimating agriculture and causing global famine.

**organisation** (noun) Group with some degree of structure. (verb) The act of organising by an organiser.

**organiser** In the context of social movements, a person who facilitates community empowerment. (S)he may do this by guiding the formation and growth of one or more climate action groups; helping a group to develop a shared vision, strategy and tactics; fostering a democratic group structure and decision-making processes; and organising public meetings, workshops, study groups and actions. There are elements of the trade union organiser or cadre in this concept.

**peak demand reduction** Reducing the height of peaks in daily electricity demand.

**peak-load** Daily peaks in electricity demand. Also describes type of power station that is used specifically for supplying peaks in demand: eg, gas turbine or hydroelectricity.

**peak oil** Peak in production of oil. Can be applied to the whole planet, a nation or an oil-field.

**permafrost** Frozen ground. Vast areas exist in Siberia, northern Canada and Alaska.

**photosynthesis** Natural process by which plants capture $CO_2$ from the atmosphere and solar energy from sunlight to form carbohydrates, which store the solar energy.

**photovoltaic cell** Semiconductor material that directly produces electricity when exposed to sunlight.

**potential energy** Energy stored in some form, such as chemical or nuclear bonds, or energy due to height above a reference level.

**power** Rate of energy conversion. Sometimes used loosely instead of 'electricity'.

**primary energy** Energy sources obtained directly from the environment, eg, coal, oil, gas, wood, hydroelectricity, wind, wave, solar. Inappropriately called 'energy production' by economists.

**pyrolysis** Heating a fuel in a limited supply of oxygen, so that it smoulders.

**ramping** Varying the output of a power station up or down.

**rated power (of power station)** Maximum or peak power output recommended by manufacturers for continuous normal operation.

**reactor-grade plutonium** Plutonium extracted by reprocessing from the spent fuel of civil nuclear reactors.

**rebound effect** Situation in which all or part of economic savings from energy efficiency are invested in using more energy. Sometimes used more generally to describe the macro-economic growth that follows from increased economic efficiency of production.

**renewable energy portfolio standard** Support mechanism for renewable energy that mandates that electricity retailers purchase a minimum proportion of their electricity from renewable energy sources. Usually in the form of a tradable renewable energy certificate scheme.

**renewable energy target** Target for renewable energy generation. May be part of a renewable portfolio standard or a tradable renewable energy certificate scheme.

**reprocessing** Chemical process to extract plutonium and unused uranium from the highly radioactive spent fuel of a nuclear reactor. Uses remote handling techniques.

**scenario study** Desktop study that explores the future consequences of different sets of assumptions.

**social justice** Equal access for all to basic human needs, such as food and water, shelter, personal security, health services and education.

**solar thermal electricity** Electricity generated by using the heat from focused sunlight to boil water to produce steam to generate electricity. Also called **concentrated solar thermal power.**

**somatic energy** Solar energy that is converted from one form to another through metabolic processes within living organisms.

**spent fuel** Highly radioactive used fuel that is removed from a nuclear reactor after a period of operation. It contains fission products (such as strontium-90 and caesium-137), some unused uranium, and transuranic elements (such as plutonium-239) created in nuclear reactions.

**spilled** or **curtailed energy** Energy dumped when supply exceeds demand.

**stationary energy** All energy production and consumption except for transport.

**stranded asset** Item of economic value, owned by an individual or corporation, that is worth less on the market than it is on a balance sheet due to the fact that it has become obsolete in advance of complete depreciation. For example, a conventional coal-fired power station after a large carbon tax has been introduced.

**strategy** Planning and conduct of long-term campaign to achieve broad goals.

**sustainable development** Best-known (Brundtland) definition is: to meet the needs of the present, without compromising the ability of future generations to meet their own needs. Another definition is: types of economic and social development that protect and enhance the environment and social equity.

**sustainable energy system** System of technologies, laws, institutions, education, industries and prices governing energy demand and supply for the sustainable development process and ultimately for achieving a sustainable society.

**synthesis gas** Mixture of carbon monoxide and hydrogen, produced by gasifying wood or coal.

**tactics** Individual steps or tools used in carrying out a strategy.

**thermal efficiency** In the process of energy conversion, useful energy output divided by energy input, usually expressed as a percentage by multiplying by 100.

**tradable emission permit** Permit to emit a specified quantity of $CO_2$ or other pollutant. The permit has monetary value and may be traded in a market.

**tradable renewable energy certificate** Tradable certificate issued in exchange for 1 MWh of renewable electricity fed into the grid. Part of a renewable energy portfolio standard.

**transaction cost** Economic cost involved in buying or selling a product that is additional to the price of the product. For example, the cost of transport to a shop, the commission on buying or selling shares, or the value of time spent in obtaining information about a product.

**transit city** City in which the majority of passenger trips are by public transport.

**transmission line** Power line for carrying large quantities of electricity over long distances at high voltage.

**transuranic element** Element heavier than uranium.

**trigeneration** Energy generation plant that supplies three forms of energy: electricity, heating (space or water) and cooling.

**upstream** At points of production or import (of fossil fuels).

**watt** Basic unit of power in SI units, the rate of change of energy generation or energy use over time. 1 watt = 1 joule per second.

**weapons-grade plutonium** Plutonium extracted from the spent fuel of a reactor that is operated in a manner to reduce the 'contamination' of Pu-239 with other isotopes (such as Pu-238) that would make it less 'efficient' as a nuclear explosive.

# ABBREVIATIONS

| AC | alternating current |
|----|---------------------|
| AETA | Australian Energy Technology Assessment |
| BAU | business-as-usual |
| c/kWh | cents per kilowatt-hour |
| CAG | climate action group |
| CCS | carbon (dioxide) capture and sequestration or storage |
| $CO_2$ | carbon dioxide |
| CPV | concentrated solar PV |
| CSIRO | Commonwealth Scientific and Industrial Research Organisation (Australia) |
| CST | concentrated solar thermal |
| DC | direct current |
| DNI | Direct Normal Incidence or Irradiance |
| EE | energy efficiency |
| EEG | Erneuerbare-Energien-Gesetz (Germany's Renewable Energy Act) |
| ENGO | environmental non-government organisation |
| EROI | energy return on investment |
| ESCO | energy service company |
| ETS | emissions trading scheme |
| EU | European Union |
| EV | electric vehicle |
| FiT | feed-in tariff |
| GDP | gross domestic product |
| GEA | Global Energy Assessment |
| GFC | Global Financial Crisis |
| GHG | greenhouse gas |

| GHI | Global Horizontal Incidence or Irradiance |
| HEV | hybrid electric vehicle |
| IAEA | International Atomic Energy Agency |
| IEA | International Energy Agency |
| IPCC | Intergovernmental Panel on Climate Change |
| IRENA | International Renewable Energy Agency |
| LNG | liquefied natural gas |
| LOLP | loss-of-load probability |
| LRET | Large-Scale Renewable Energy Target (Australia) |
| NEM | National Electricity Market (Australia) |
| NGO | non-government organisation |
| NIMBY | Not in My Back Yard |
| OECD | Organisation for Economic Co-operation and Development |
| PHEV | plug-in hybrid electric vehicle |
| PV | photovoltaic |
| R&D | research and development |
| RE | renewable energy |
| REC | renewable energy certificate |
| RElec | renewable electricity |
| RGGI | Regional Greenhouse Gas Initiative |
| RPS | Renewable (energy) Portfolio Standard |
| SRES | Small-Scale Renewable Energy Scheme (Australia) |
| UNFCCC | United Nations Framework Convention on Climate Change |
| UNSW | University of New South Wales |

# UNITS AND CONVERSION FACTORS

### Powers of 10

| Prefix | Symbol | Value | Example |
|---|---|---|---|
| kilo | k | $10^3$ | kilowatt kW |
| mega | M | $10^6$ | megawatt MW |
| giga | G | $10^9$ | gigajoule GJ |
| tera | T | $10^{12}$ | terawatt-hour TWh |
| peta | P | $10^{15}$ | petajoule PJ |
| exa | E | $10^{18}$ | exajoule EJ |

### SI units

| Basic unit | Name | Symbol |
|---|---|---|
| length | metre | m |
| mass | kilogram | kg |
| time | second | s |
| temperature | degree Kelvin | °K |

| Derived unit | Name | Symbol |
|---|---|---|
| energy | joule | J |
| power | watt | W |
| potential difference | volt | V |
| temperature | degree Celsius | °C |
| time | hour | h |

## Conversion factors

| Type | Name | Symbol | Value |
|------|------|--------|-------|
| volume | cubic metre | $m^3$ | 1000 litres = 1 kL |
| volume | gallon (US) | | 3.785 L |
| energy | kilowatt-hour | kWh | $3.6 \times 10^6$ J = 3.6 MJ |
| energy | terawatt-hour | TWh | $3.6 \times 10^{15}$ J = 3.6 PJ |
| energy | litre of petrol | L | $3.2 \times 10^7$ J |
| energy | barrel of oil | | $6.12 \times 10^9$ J = 6.12 GJ |
| energy | barrel of oil | | 159 L |
| energy | British thermal unit | btu | 1055.1 J = 1.0551 kJ |
| energy | cubic metre of natural gas at STP* | $m^3$ | $3.4 \times 10^7$ J |
| energy | tonne of black coal | t | 23 GJ |
| energy | tonne of brown coal | t | 10 GJ |
| energy | tonne of green wood | t | 10 GJ |
| energy | tonne of oven-dried wood | t | 20 GJ |
| carbon | tonne of carbon in $CO_2$ | t | 3.67 t $CO_2$ |
| power | kWh per year | kWh/y | 0.114 W |
| time | year | y | 8760 hours |
| temperature | degree Celsius | °C | (°C + 273) °K |
| speed | kilometre per hour | km/h | 0.2778 m/s |

*STP is 'standard temperature & pressure', defined as 1 atmosphere of pressure and 0°C.

# FURTHER READING

## Greenhouse science

Hansen, J, Sato, M, Kharecha, P, Beerling, D, Berner, R, Masson-Delmotte, V, Pagani, M, Raymo, M, Royer, DL and Zachos, JC (2008) Target atmospheric $CO_2$: Where should humanity aim?, *Open Atmospheric Science Journal*, 2: 217–31.

Intergovernmental Panel on Climate Change (IPCC) (2013–14) *AR5: Fifth Assessment Report*, IPCC, Working Group reports: WGI: The Physical Science Basis; WGII: Impacts, Vulnerability and Adaptation; WGIII: Mitigation of Climate Change, <www.ipcc.ch>.

McKibben, B (2012) Global warming's terrifying new math, *Rolling Stone*, 19 July <www.rollingstone.com/politics/news/global-warmings-terrifying-new-math-20120719>; see also <http://math.350.org>.

Potsdam Institute for Climate Impact Research and Climate Analytics (2012) *Turn Down the Heat: Why a 4°C warmer world should be avoided*, World Bank, Washington DC.

Real Climate <www.realclimate.org> – a website by climatologists.

Schiermeier, Q (2010) The real holes in climate science, *Nature*, 21 January, 463: 284–87.

Skeptical Science: Getting skeptical about global warming skepticism <www.skepticalscience.com>.

## Climate mitigation in general

Climate Spectator <www.climatespectator.com.au> – website and bulletin, edited by Tristan Edis.

Delina, L and Diesendorf, M (2013) Is wartime mobilisation a suitable policy model for rapid national climate mitigation?, *Energy Policy*, 58: 371–80.

GEA (2012) *Global Energy Assessment – Toward a Sustainable Future*, Cambridge University Press, Cambridge, and the International Institute for Applied Systems Analysis, Laxenburg, Austria, <www.globalenergyassessment.org>.

Hamilton, C (2013) *Earthmasters,* Yale University Press and Allen & Unwin, Sydney.

RenewEconomy <http://reneweconomy.com.au> – website and bulletin, edited by Giles Parkinson.

## Renewable energy

Boyle, G (ed) (2012) *Renewable Energy: Power for a sustainable future*, 3rd ed, Oxford University Press, Oxford.

Delucchi, MA and Jacobson, MZ (2011) Providing all global energy with wind, water, and solar power, Part II: Reliability, system and transmission costs, and policies, *Energy Policy*, 39: 1170–90.

Elliston, B, MacGill, I and Diesendorf, M (2013) Least cost 100% renewable electricity scenarios in the Australian National Electricity Market, *Energy Policy*, 59: 270–82.

European Wind Energy Association <www.ewea.org>.

Eurosolar <www.eurosolar.de>.

International Network for Sustainable Energy (INFORSE) <www.inforse. org>.

International Renewable Energy Agency (IRENA) <www.irena.org>. International Solar Energy Society (ISES) <www.ises.org>.

Jacobson, MZ and Delucchi, MA (2011) Providing all global energy with wind, water, and solar power, part I: Technologies, energy resources, quantities and areas of infrastructure, and materials, *Energy Policy*, 39(3): 1154–69.

Kemp, M (ed) (2010) *Zero Carbon Britain 2030: A new energy strategy*, Centre for Alternative Technology, <www.zerocarbonbritain.org>.

National Renewable Energy Laboratory (NREL) (2012) *Renewable Electricity Futures Study*, Technical Report TP-6A20-52409, National Renewable Energy Laboratory, Golden, Colorado, <www.nrel.gov/analysis/re_ futures>.

Renewable Energy Policy Network for the 21st Century (REN21) (2013) *Renewables 2013 Global Status Report*, <www.ren21.net>.

Sørensen, B (2010) *Renewable Energy: Its physics, engineering, environmental impacts, economics and planning*, 4th ed, Academic Press, San Diego – textbook suitable for scientists and engineers.

## Energy efficiency

Pears, A (2007) Imagining Australia's energy services futures, *Futures*, 39: 253–71.

## Transport and urban form

Newman, P (2008) *Cities as Sustainable Ecosystems*, Island Press, Washington DC.

Newman, P, Beatley, T and Boyer, H (2009) *Resilient Cities: Responding to peak oil and climate change,* Island Press, Washington DC.

Newman, P and Kenworthy, J (1999) *Sustainability and Cities: Overcoming automobile dependence*, Island Press, Washington DC.

Victoria Transport Policy Institute, British Columbia, Canada, <www.vtpi. org>.

White, D, Sutton, P, Pears, A, Dick, J and Crow, M (1978) *Seeds for Change: Creatively confronting the energy crisis*, Conservation Council of Victoria and Patchwork Press, Melbourne.

## Nuclear energy

Institute for Science and International Security (ISIS) <http://isis-online.org>.
Nuclear Weapons Archive <http://nuclearweaponarchive.org>.
Schneider, M and Froggatt, A (2013) *World Nuclear Industry Status Report 2013*,
    Mycle Schneider Consulting, Paris, July, <www.worldnuclearreport.org>.
Sovacool, BK (2011) *Contesting the Future of Nuclear Power,* World Scientific,
    New Jersey.

## Economics, markets and carbon pricing

Australian Broadcasting Corporation (ABC) Radio National, Rear Vision,
    (2011), Carbon Tax, 12 January, <www.abc.net.au/rn/rearvision/
    stories/2011/3085312.htm#transcript>.
Center for the Advancement of the Steady State Economy <http://steadystate.
    org>.
Dietz, R and O'Neill, D (2013) *Enough is Enough: Building a sustainable
    economy in a world of finite resources*, BK, San Francisco.
European Commission, Climate Action, <http://ec.europa.eu/clima/policies/
    ets/index_en.htm>.
Jackson, T (2009) *Prosperity without Growth: Economics for a finite planet*,
    Earthscan, London.
Spash, C (2010) The brave new world of carbon trading, *New Political
    Economy*, 15(2): 169–95.

## Barriers to, myths about and politics of mitigation

ABC (2006) *The Greenhouse Mafia,* Four Corners, Program transcript, 13
    February, <www.abc.net.au/4corners/content/2006/s1568867.htm>.
Beder, S (2000) *Global Spin: The Corporate Assault on Environmentalism*, Revised
    ed, Scribe, Melbourne.
Keane, S (2011) The ugly landscape of the Guardians, *Independent Australia*,
    <www.independentaustralia.net/2011/environment/the-ugly-landscape-of-
    the-guardians>.
Klein, N (2011) Capitalism versus the climate, *The Nation*, <www.thenation.
    com/article/164497/capitalism-vs-climate?page=0,0>.
Oreskes, N and Conway, E (2010) *Merchants of Doubt: How a handful of
    scientists obscured the truth on issues from tobacco smoke to global warming,*
    Bloomsbury, London.
Pearse, G (2007) *High and Dry*, Viking, Melbourne. See also <www.guypearse.
    com>.
Washington, H and Cook, J (2011) *Climate Change Denial: Heads in the sand*,
    Routledge-Earthscan, London.

## Community climate action

100% Renewable Energy <www.100-percent.org>.
350.org <http://350.org>.
Australian Youth Climate Coalition <http://aycc.org.au>.
Bill McKibben <www.billmckibben.com>.

Climate Action Network <www.climatenetwork.org>.

Climate Action Network Australia <http://cana.net.au>.

Ekins, P (1992) *A New World Order: Grassroots movements for global change*, Routledge, London.

Right Livelihood Awards <www.rightlivelihood.org>.

## Strategies for social change movements

Alinsky, S (1971) *Rules for Radicals: A practical primer for realistic radicals*, Random House, New York.

Bobo, K, Kendall, J and Max, S (2001) *Organising for Social Change: Midwest Academy Manual for Activists*, 4th ed, The Forum Press, Santa Ana CA.

Coover, V, Deacon, E, Esser, C and Moore, C (1978) *Resource Manual for a Living Revolution*, 2nd ed, New Society Press, Philadelphia PA.

Diesendorf, M (2009) *Climate Action: A campaign manual for greenhouse solutions*, UNSW Press, Sydney.

Gandhi, MK (1927 & 1929) *An Autobiography or The Story of my Experiments with Truth*, Navajivan Publishing House, Ahmedabad.

Lakey, G (1973) *Strategy for a Living Revolution*, WH Freeman, Boston.

Moyer, B, McAllister, J, Finley, ML and Soifer, S (2001) *Doing Democracy: The MAP model for organising social movements*, New Society Publishers, Gabriola BC, Canada.

Rose, C (2005) *How to Win Campaigns: 100 steps to success*, Earthscan, London.

Sharp, G (1973) *The Politics of Nonviolent Action*, Porter Sargent Publisher, Boston MA.

# LIST OF TABLES, FIGURES AND BOXES

## Tables

## Figures

## Boxes

# INDEX